D0186471

# *Home*
# **Plumbing**

# Home
# **Plumbing**

GUILD OF
MASTER CRAFTSMAN
PUBLICATIONS

## **Phil Thane**

First published 2009 by
**Guild of Master Craftsman Publications Ltd**
Castle Place, 166 High Street,
Lewes, East Sussex BN7 1XU

Text © Phil Thane 2009
© in the Work GMC Publications Ltd, 2009

ISBN 978-1-86108-649-5

All rights reserved

The right of Phil Thane to be identified as the author of this work has
been asserted in accordance with the Copyright, Designs and Patents
Act 1988, sections 77 and 78.

No part of this publication may be reproduced, stored in a retrieval
system or transmitted in any form or by any means without the prior
permission of the publisher and copyright owner.

This book is sold subject to the condition that all designs are
copyright and are not for commercial reproduction without the
permission of the designer and copyright owner.

The publishers and author can accept no legal responsibility for any
consequences arising from the application of information, advice or
instructions given in this publication.

A catalogue record for this book is available from the British Library.

**Associate Publisher** Jonathan Bailey
**Production Manager** Jim Bulley
**Managing Editor** Gerrie Purcell
**Editors** Polly Goodman/Alison Howard
**Managing Art Editor** Gilda Pacitti
**Design** JC Lanaway

**Photography and illustration:** Phil Thane

Set in Clarendon and MetaPlus
Colour origination by GMC Reprographics
Printed and bound in Thailand by Kyodo Nation Printing

# Contents

# 4

## Bathrooms, Showers, Kitchens and Heating

pages 78–111

## Troubleshooting

pages 112–119

# 5

# Introduction

## CAUTION
### BEFORE YOU START

■ Before starting a project, read the instructions carefully and follow any guidelines supplied with equipment or materials used. Where necessary, seek professional advice.

■ Many aspects of plumbing work and installations are subject to building regulations and/or local bye-laws. Check with your local authority and regional water supplier to ensure that the work you are planning meets all relevant regulations and requirements.

■ Follow accepted safety procedures when carrying out work on your home. If in any doubt, seek professional advice before starting.

It is no coincidence that planning authorities consider the provision of water supply and drainage to be the critical factor that turns a building into a dwelling. Without them, a building is no more than a glorified shed; with them it can be a home. An efficient and reliable domestic water system is fundamental to the enjoyment of a home; a well-planned system in good order enhances the lives of the occupants, but an unreliable, inadequate or inconvenient system is a constant irritation.

Considering this, it's no surprise that the value of a house is significantly affected by the quality of its plumbing system and bathroom facilities. Upgrading bathroom sanitary ware and fittings, adding en suite facilities to a master bedroom, or even installing a shower unit above a bath, usually adds far more to the value of a property than the cost of the work.

Though at one time the equipment, materials and fittings used in household water systems were the province of specialist suppliers, these days they are readily available to the householder at affordable prices – and new, user-friendly products and technologies have made them easier than ever to install. The range can be bewildering and modern water systems can be complicated. This book aims to give the householder a good overview of how plumbing works, the range of products available, which of them to choose, and how to undertake basic installations and maintenance.

## CAUTION
### GAS PLUMBING

■ Although gas plumbing may appear similar to that used for water supply, it is highly specialised and mistakes can result in serious injury, loss of life and/or significant property damage.

■ On no account must ANY work on gas supplies, installations or appliances be attempted by anyone other than a Gas Safe-registered technician.

SAINTE DEN STOC

WITHDRAWN FROM
DÚN LAOGHAIRE-RATHDOWN COUNTY
LIBRARY STOCK

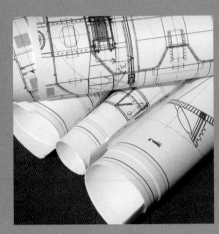

# 1

# Knowledge
# and the Basics

Domestic plumbing work is usually quite

simple, once you know how things work,

what they are called, where they fit, and why.

All this is covered in Chapter One...

# Supply

**SUPPLY AND WASTE**
Supply and overflow
plumbing in a
bathroom (*above*).

**PIPE LINES**
Waste pipes are much
larger than supply
pipes (*right*).

A household water system should be
viewed as having two distinct parts:

- Supply, including all pipework and fittings
  that deliver water from the mains water
  supply to the point of use, and
- Waste, including all pipework and fittings
  that take used water from the point of use
  to the drains.

Water on the supply side of a system is
under pressure, so its pipework is of a small
diameter – 15 and 22mm are standard sizes
for supply pipes – and joints and fittings are
designed to withstand the force of the
pressurised water they contain.

Waste water moves through the system
by gravity only, so neither waste pipes nor
fittings need be pressure-resistant. However,
for blockages to be avoided the water must

be able to flow freely, so waste pipes are of a large diameter and installed at a sufficient 'fall' (angle from the horizontal) for the water to drain quickly away. A fall of about 20mm in every metre is sufficient.

Fresh water is supplied by regional water companies via large-diameter pipes called water mains that are usually buried beside or under roads. Where the main passes the house, a small-diameter pipe known as the communication pipe branches off to carry water to the company stopcock.

The company stopcock is a control valve and is usually found under a small metal cover near the boundary of the property, set into the pavement, driveway or garden. It is approximately 1m below the surface, and is turned on and off using a long key. In an emergency it may be possible to use a piece of wood with a notch cut into its end to do the same job. Make sure that you know where the stopcock for your house is located.

At the company stopcock, the water company's responsibility ends and the householder's begins, so the service pipe that carries water from the stopcock into the property is part of the household water system. It must be buried between 750mm and 1.35m below ground to avoid frost damage, and should not be stretched tight or ground movement may strain the joints at either end.

The service pipe usually enters the house under the kitchen sink, at which point it is known as the 'rising main'. In properties that are older or have been remodelled it may do so elsewhere. At this point a second stopcock should be provided, although in older houses this might be missing. It is useful to have a drain cock immediately above this stopcock to allow the household system to be drained (see *Repair or Replace a Stopcock*, page 73).

What happens next depends on the type of system in place. In the UK, two are in common use: indirect and direct.

**YOUR STOPCOCK**
In most houses the stopcock is on an internal wall, usually under the kitchen sink (*below left*).

**COMPANY STOPCOCK**
The company stopcock (*below*) is usually found underground near the boundary of your property.

## Anatomy of a **Company Stopcock**

**1** Water meter
**2** Stopcock

**DIRECT SUPPLY**
The kitchen tap is always supplied direct from the rising main (*above*).

---

**SAFE WATER**
Modern covered cisterns ensure potable water, even from indirect supplies (*right*).

---

## Indirect Systems

If your house has an indirect system, only the cold tap fitted to the kitchen sink draws water directly from the rising main, though kitchen appliances such as dishwashers and washing machines, or an outside tap, may sometimes be supplied directly. The rising main then travels up through the house to supply the main cistern, or water tank, usually installed in the roof space. All other cold water in the house is supplied under gravity pressure from this cistern.

There are several advantages of such an indirect system. Most of the pipework and fittings in the house are under the pressure of gravity only, and are therefore less likely to leak. Should a leak occur, less water will escape, therefore reducing damage to the house and its contents. Low–pressure water is quieter as it moves through the system, which is especially true of WC cisterns, and puts less strain on valves and washers. Water drawn from a household cistern is usually warmer than from the main, so it requires less energy to heat, and reduces

condensation on pipework and lavatory cisterns. In addition, if the mains supply is temporarily interrupted, a cistern-full of water is still available for essential use.

The chief disadvantage of indirect systems is their complexity. Taking the mains supply to the roof space, storing it in a cistern, then redistributing it through the house requires a large quantity of pipework, as well as the installation of the cistern itself. This means higher installation costs and a higher risk of frozen pipes in a cold roof space which may lead to leaks.

## Potable Water Regulations

Until recently, building regulations required only one tap in a house, usually in the kitchen, to be supplied with drinking water direct from the mains. New regulations now stipulate that storage cisterns should also supply drinkable ('potable') water. These regulations are known as Bye-law 30 in England and Wales and as Bye-law 60 in Scotland. Though they vary slightly, the object is to keep, dust, dirt, insects and light out of the cistern. If your cistern is fairly modern with a close-fitting lid you can achieve this by fitting a Bye-law 30 (or 60) kit. If you have an old cistern you can replace it, but depending on the age of the boiler and hot water cylinder, it may make more sense to convert to a direct system.

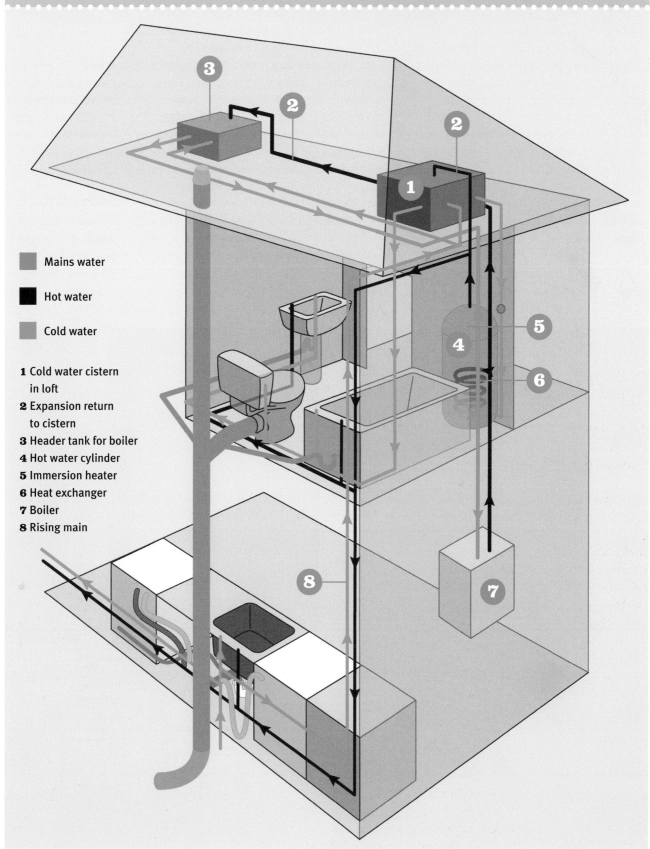

Mains water

Hot water

Cold water

**1** Cold water cistern
in loft
**2** Expansion return
to cistern
**3** Header tank for boiler
**4** Hot water cylinder
**5** Immersion heater
**6** Heat exchanger
**7** Boiler
**8** Rising main

# Direct Systems

Indirect systems are not used outside the UK, and direct systems are standard in the rest of Europe. They are becoming more popular in the UK because they are cheaper to install, save space and easily meet new regulations. Compare the illustration opposite, on page 17, with that on page 15 to see how much simpler it is. If your house is quite modern, or it has had a complete replacement system installed in it recently, you may have a direct system.

In a direct cold-water system, all taps, cisterns and appliances draw water directly from the mains supply, with no need for a storage cistern in the roof space (unless one is required for the hot water or heating system). The advantages are that pipework is kept to a minimum, reducing both the cost of installation and the opportunities for problems, and as all cold taps are supplied from the mains, they supply potable water and therefore satisfy the regulations.

One disadvantage is that all cold water pipework and fittings in your house must be able to resist mains pressure. Leaks are more likely and any that do occur will be more severe as the mains pressure will force greater amounts of water through the leak. A properly installed and maintained system should not present problems. Most people with direct systems notice only that they are noisier, especially WC cisterns when they are refilling, and that the water is colder, which results in more condensation on pipes and cisterns. This seems a fair price to pay for a simpler, lower-cost system.

**DIRECT SYSTEM**
A direct plumbing system (*right*) is much more straightforward than an indirect one (*previous page*).

**KITCHEN STYLE**
A beautiful and well-designed kitchen can become the heart of the home (*below*).

# Anatomy of a **Direct Plumbing System**

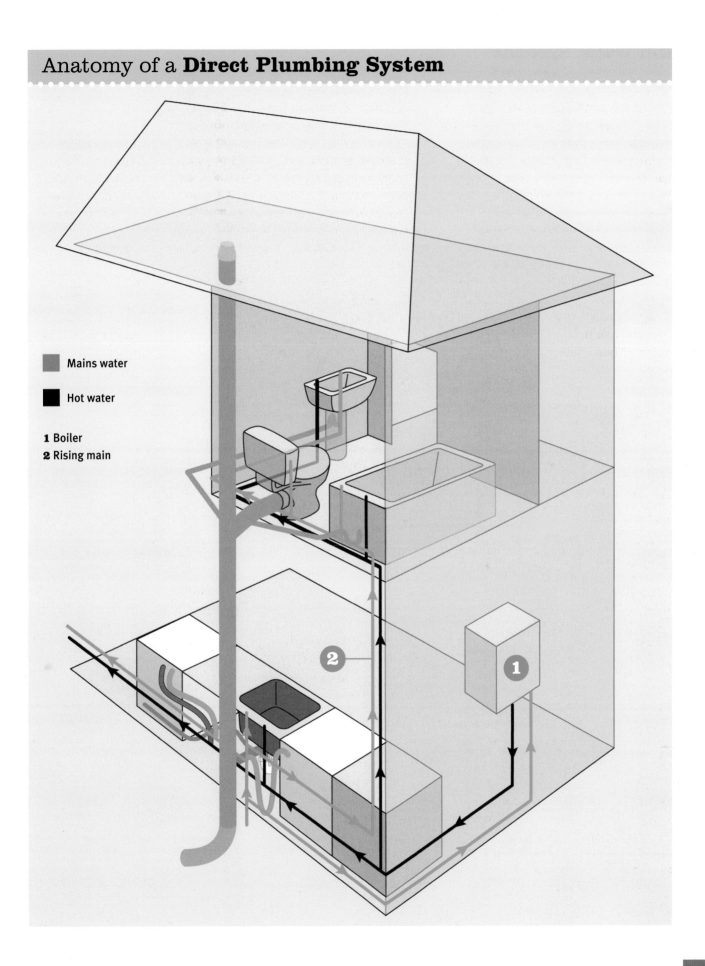

Mains water

Hot water

**1** Boiler
**2** Rising main

# Hot Water Systems

There are several different ways to provide hot water and central heating. Traditionally, the hot taps are supplied from a hot water storage cylinder, which in turn draws from a cold water storage cistern in the roof space. Water is heated in the cylinder by a fitted immersion heater and/or indirectly by a boiler. Hot water is drawn as needed from that stored in the cylinder.

Older houses and flat conversions sometimes have instantaneous gas multipoint heaters, fed directly from the rising main, which only light when a hot tap is opened. There is no energy lost from a storage cylinder, so these systems are quite efficient when a substantial amount of water is drawn off. They are not so good, however, when used in quick bursts, for example to wash hands or rinse a cup. As most people now want central heating as well as hot water they have been largely superseded by combi (combination) boilers.

## Anatomy of a **Cylinder System**

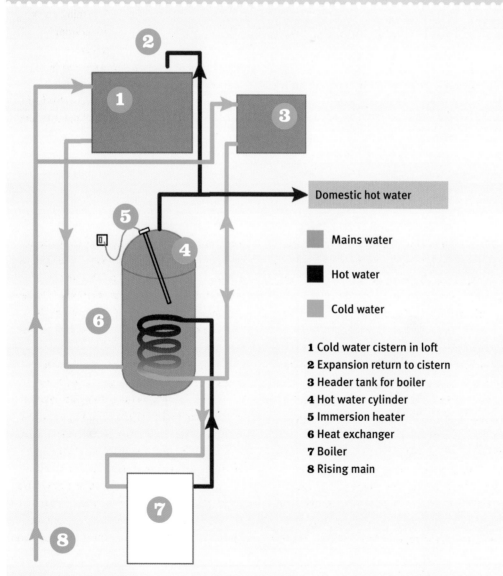

Domestic hot water

Mains water

Hot water

Cold water

1 Cold water cistern in loft
2 Expansion return to cistern
3 Header tank for boiler
4 Hot water cylinder
5 Immersion heater
6 Heat exchanger
7 Boiler
8 Rising main

# Anatomy of a **Combi System**

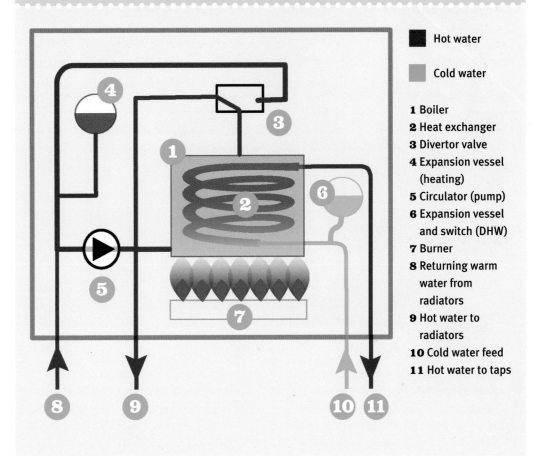

**Hot water**

**Cold water**

1 Boiler
2 Heat exchanger
3 Divertor valve
4 Expansion vessel
  (heating)
5 Circulator (pump)
6 Expansion vessel
  and switch (DHW)
7 Burner
8 Returning warm
  water from
  radiators
9 Hot water to
  radiators
10 Cold water feed
11 Hot water to taps

Combi (combination) boilers are increasingly popular, especially in flats and small houses where space is limited, as they cut out the need for a hot water cylinder. They are supplied from the rising main so no storage cistern is needed, and they provide water heating and central heating in a single unit.

The simplified drawing of a combi boiler explains how this type of boiler works. There are two circuits, so the dirty water that circulates through the radiators does not mix with the domestic hot water (DHW) going to the taps. A divertor valve switches heat from radiators to DHW as required.

The illustration above shows the valve sending hot water out to the radiators. The returning cooler water goes back into the boiler for reheating and the circulator keeps it all moving. When a tap is opened, cold water runs through the heat exchanger and is heated, but as it does so the temperature of the boiler falls and a thermostat triggers the divertor valve so hot water from the boiler is sent back into it instead of to the radiators.

A boiler like this will supply hot water quickly when the heating is already on. When the heating is off a different system comes into action. Opening a tap causes a pressure drop in the pressure vessel. This operates a switch, which fires the boiler. This takes a few seconds longer and to counter that, some boilers have a small hot water storage vessel built in ready from immediate use. Combi boilers are usually gas fired but there are oil-fired versions too.

# Waste

The majority of UK dwellings discharge their waste water into main sewers that are maintained by the same regional water company that provides mains water supplies. Outside built-up areas, where main sewers are not present, households will have their own septic tanks or treatment plants. In either case, the requirements for pipework directly servicing the house will be the same. In much the same way as the householder's involvement with water supply starts at the main stopcock, it may be considered that the household waste water system ends at the inspection chamber that is situated near the property under an airtight metal cover.

The inspection chamber is the point where one or more drains from the house merge to form the main drain, which discharges into the sewer or septic tank. The drain or drains run underground to the house itself from this inspection chamber and, should a blockage occur, it must be dealt with from here.

Where waste water from the house enters the outside drains, two different types of system are in use: two-pipe and single-stack. The former was largely superseded by the single-stack system in the 1960s, so older houses are likely to employ the two-pipe system and newer houses the single-stack system. Having said this, older houses that have been extended or modernised might well use a combination of the two.

## Two-pipe waste drainage

In a two-pipe system, waste from WCs on the first floor and higher is carried to outside drains by a dedicated large-diameter pipe known as a soil pipe, which is fixed to the outside wall of the house. This extends upwards to just above the eaves, where it allows gases to vent, and is covered with a basket-like guard. Ground-floor WCs usually discharge directly into the underground drain. A second pipe, of smaller diameter than the soil pipe and referred to as the waste pipe, is also fixed to the outside wall. Into this is discharged the waste water from sinks, basins, baths and showers on the first floor and higher. At ground level the waste pipe discharges into a gully – an open drain covered with a grille – as do individual waste pipes from ground floor sinks and so on.

**MAIN DRAIN**
Your responsibility ends where your drain meets the public sewer – the inspection chamber (*above*).

# Single-stack waste drainage

In a single-stack system, all waste water from WCs, sinks, basins, baths, showers and kitchen appliances is conveyed to the underground drains by a single, large-diameter pipe known as a 'stack'. Where a stack has been built into the house it may be inside an exterior wall, or may even be fitted in a central duct inside the house. Like the soil pipe of a two-pipe system, the top of the stack projects above the eaves to allow the release of gases, and is covered by a guard resembling a basket.

In some installations the stack terminates inside the building and must be capped with an air admittance valve to prevent siphoning and be contained within a ventilated duct or roof space. The regulations concerning waste stacks and valves are complex, so professional advice must be sought before making any alterations to such a system (see *Blocked Waste Pipes* and *Unblock a Waste Pipe* on pages 116–19 for advice on working on wastewater systems).

**TWO SYSTEMS**

In the two-pipe system (*below left*), the bath and basin wastes enter a hopper to the waste pipe (shared with the rainwater downspout) and the WC stack is separate. In a single-stack system (*below*) the bath, shower and basin waste pipes empty into the same soil stack as the WC.

# Basic Tools

You do not need a vast collection of specialised tools for plumbing work. Many are also used for other DIY tasks, so they will be a good investment. It is useful to have some basics to hand to deal with emergencies, so these are listed first. You can put off buying the specialised tools (see page 26) until you really need them for an advanced project.

## Drain Clearing Liquid

The most common plumbing problem facing householders is a blocked waste pipe. It's nasty, smelly, and the rest of the household will want you to deal with NOW! You may not consider drain-clearing chemicals 'tools',

but they can save you some unpleasantness, so keep some handy. Essentially these are powerful acid or alkaline solutions that will dissolve most of the things likely to block a drain, and there are various trade names. Needless to say anything that is powerful is poisonous and very bad for skin, eyes, mouths and noses. Keep them well away from children and pets.

Drain-clearing liquids can only operate where they can reach. If a long section of pipe is clogged, the liquid will have become neutralised (worn itself out) before it gets right through. Then it will be time for something more serious.

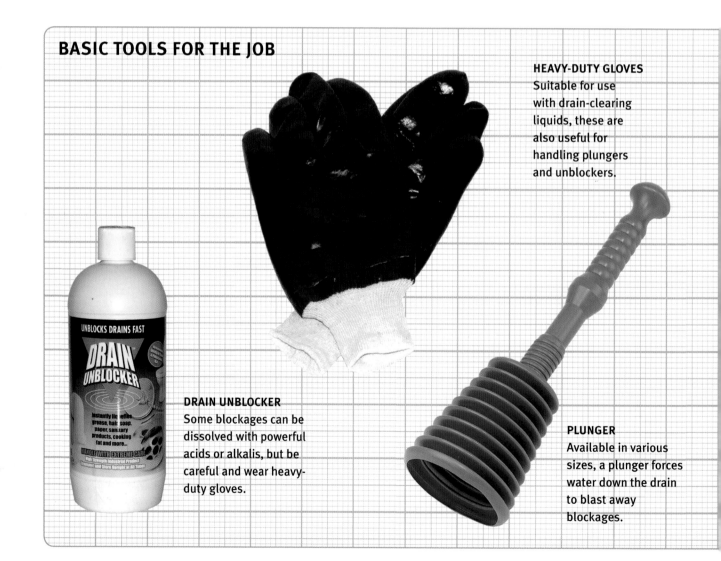

## BASIC TOOLS FOR THE JOB

**HEAVY-DUTY GLOVES**
Suitable for use with drain-clearing liquids, these are also useful for handling plungers and unblockers.

**DRAIN UNBLOCKER**
Some blockages can be dissolved with powerful acids or alkalis, but be careful and wear heavy-duty gloves.

**PLUNGER**
Available in various sizes, a plunger forces water down the drain to blast away blockages.

## Plungers

For sinks, basins, baths and showers you need a plunger. The traditional type is a rubber cup with a handle, but there are concertina types and even powered versions. To use a plunger you need some water covering the plughole, but you will probably have that already. The plunger forms a seal around the plughole and as you push it up and down, it pushes water in and out of the pipe, which is usually enough to dislodge the blockage.

## Rods and Wires

Drain rods are not something the average householder needs to worry about; they are used in underground drains to clear serious blockages. For waste pipes inside the house a corkscrew cable device may be useful. These have a spiral cable, like a curtain wire, with a corkscrew-like end to break up the blockage and a winder so you can twist it.

## Heavy-Duty Gloves

Heavy-duty waterproof gloves will protect your skin from caustic drain-clearing liquids, as well as anything else you might find down the drain.

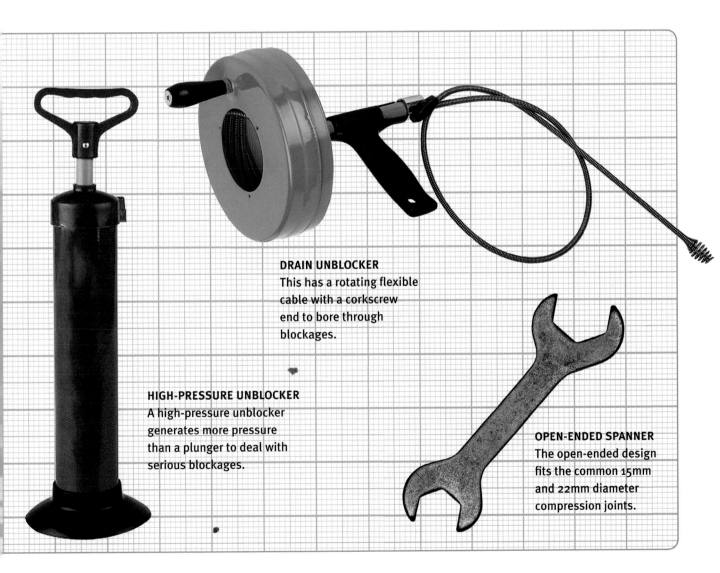

**DRAIN UNBLOCKER**
This has a rotating flexible cable with a corkscrew end to bore through blockages.

**HIGH-PRESSURE UNBLOCKER**
A high-pressure unblocker generates more pressure than a plunger to deal with serious blockages.

**OPEN-ENDED SPANNER**
The open-ended design fits the common 15mm and 22mm diameter compression joints.

## Spanner or wrench?

In the UK, a spanner is designed to turn nuts or screws with square or hexagonal heads. Wrenches are designed to grip a wide range of things – pipes and bars as well as nuts and screws. In the US, 'wrench' covers the whole range. The US usage is becoming more common in the UK too, especially for adjustable spanners.

## Spanners

Open-ended spanners have parallel jaws to fit the flats of conventional nuts. They are very useful on compression fittings used in plumbing, too. Two in the same size are generally needed to deal with compression joints, and mechanics' spanner sets usually have only one of each size. The cheapest solution is to purchase a pair of special compression joint spanners (shown below), which are double-ended to suit the most common 15mm and 22mm fittings.

Ring spanners grip the corners of nuts, but are no use for pipework because you can't get them off once the pipe is joined!

**CAUTION**
A ring spanner (*right*) is of no use in plumbing because you can't get it off the pipe!

Mechanics do not like adjustable spanners; they tend to spread and lose their grip if you put too much strain on them. But plumbing joints should never be that tight anyway, and adjustable spanners are ideal for dealing with a variety of joints.

## Wrenches

The most famous wrench is a locking 'Mole' wrench (see A, below). It is of limited use in plumbing because its flat jaws tend to crush pipes and they are not parallel, so they don't fit nuts very well. It is handy sometimes to be able to lock it on to something when there isn't much room to work.

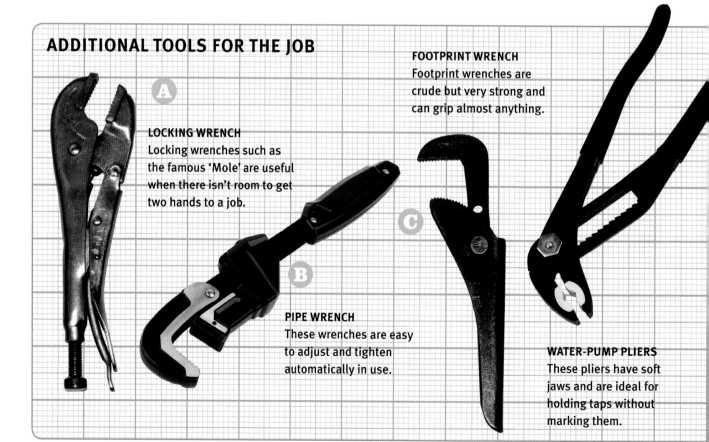

## ADDITIONAL TOOLS FOR THE JOB

**A**

**LOCKING WRENCH**
Locking wrenches such as the famous 'Mole' are useful when there isn't room to get two hands to a job.

**B**

**PIPE WRENCH**
These wrenches are easy to adjust and tighten automatically in use.

**FOOTPRINT WRENCH**
Footprint wrenches are crude but very strong and can grip almost anything.

**C**

**WATER-PUMP PLIERS**
These pliers have soft jaws and are ideal for holding taps without marking them.

There are two types of adjustable pipe wrench often referred to by common trade names Stilson (B) and Footprint (C), though various makes are available. The former, a pipe wrench, has a sliding jaw which may be controlled by a screw or a quick lock mechanism; the latter is more crude and is adjusted roughly to size by removing and replacing a thumbscrew, then tightened by squeezing the handles.

Both types are arranged so that the harder you pull on them the tighter they grip, but this only works in one direction; used the wrong way the slacken instead. If that happens, just turn the wrench over.

Adjustable pipe wrenches are useful for undoing stubborn old fittings but their power combined with serrated jaws can damage new fittings, so use them with care.

## Pliers

Plumbing (or water-pump) pliers have sliding jaws to grip a wide range of objects, including compression fittings and joints on central heating circulators. Special, soft-jaw versions can be used to hold taps while you tighten the nut underneath without marking the plated finish on the tap body.

## Screwdrivers

You will need a selection of screwdrivers for operating isolating valves, and fixing pipe clips and radiators to walls. If you are planning to do that kind of work, you will probably also need a power drill for drilling into brick or block walls.

## Anything else?

You will need safety glasses if you are drilling or soldering, a tape measure, a spirit level, a try square and a pencil. A hosepipe is essential if you have to drain a system, not to mention a bucket for emptying traps or radiators. Even better is a wide but shallow tray that will fit below a radiator valve and lots of old towels for soaking up the inevitable spills.

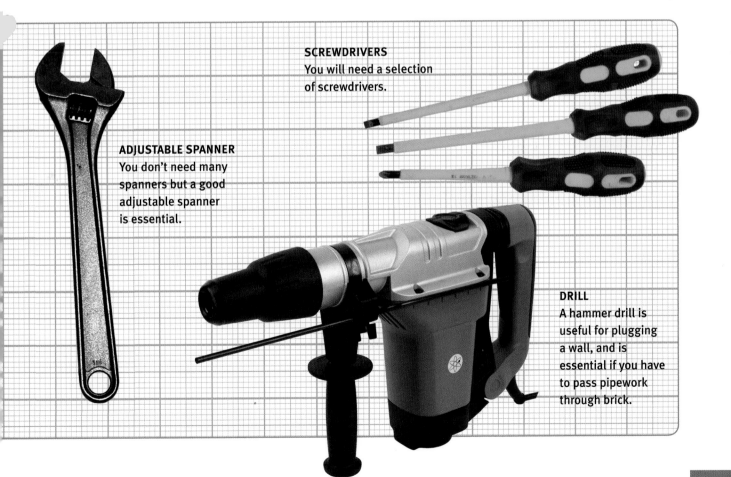

**SCREWDRIVERS**
You will need a selection of screwdrivers.

**ADJUSTABLE SPANNER**
You don't need many spanners but a good adjustable spanner is essential.

**DRILL**
A hammer drill is useful for plugging a wall, and is essential if you have to pass pipework through brick.

# Specialist Tools

## PIPE CUTTERS

You can cut the occasional piece of pipe with a hacksaw and carefully file the end smooth, but it is hard to hold a pipe without crushing it and very difficult to saw straight across it. A pipe cutter does a better job much faster.

**1** There are different designs of pipe cutters, but the principle is much the same. There are a couple of rollers to support the pipe and a sharp wheel to cut it.

**2** You fit the cutter on the pipe and twist it round.

**CUTTER WITH SCREW**
Some pipe cutters have a screw that you tighten after each rotation until the wheel has cut right through. Others tighten themselves as you twist. The former may be used on a range of pipe sizes, the latter come in different sizes.

### PVC PIPE CUTTER

Plastic pipework is quicker and easier than copper, but you do need to have the ends cut very neatly to avoid leaks. A special PVC pipe cutter like this one is a must for this sort of work.

### PIPE DEBURRER

If you are using push-fit fittings on copper pipe it is essential the pipe is smooth and the end chamfered to avoid damaging the O-ring. A deburring tool like this makes it a lot easier to do.

# PIPE BENDERS

### PIPE BENDING SPRING

Pipes kink if you try to bend them. Pipe bending springs are put inside the pipe to support it while you make the bend, then they are pulled out. They can be used on copper or plastic pipe. Springs are not necessary for occasional small jobs – you can use 'elbows' to change the direction of a pipe.

### PIPE BENDER

This can make a tighter, neater bend than a spring, but it is expensive, so it is only worth buying if you are planning to do a lot of plumbing.

## TAP RESEATING TOOL

Ceramic disk taps have largely replaced the old tap washer type, but if you have an old-style tap that drips even when you have replaced the washer, you need a tap reseating tool.

# SOLDERING

Plastic plumbing is becoming more common, but copper is still popular. Copper pipes with soldered joints take up less space than plastic, or compression joints on copper, so you may want to learn how to solder. If you do, you will need:

**CLEANING PADS**
These are essential to ensure that pipes are free from dirt or grease.

**PROPANE BLOWTORCH**
This is suitable only for soldering and cannot be used for welding.

**LEAD-FREE SOLDER**
The use of lead-free solder has been obligatory since 2006.

**FLUX**
Plumbing flux keeps the copper clean and helps the solder flow.

# JOINTING

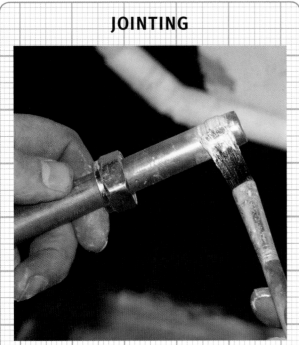

## JOINTING COMPOUND

If you are using compression joints on copper pipe (see page 43), a little jointing compound will help to seal the joint.

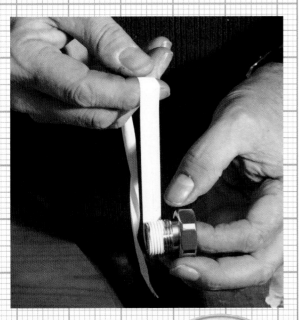

## PTFE TAPE

This tape is used to ensure a good seal in joints that screw together, like those on a tap.

# SPECIALISED SPANNERS

## RADIATOR KEY

For bleeding air out of central heating systems a radiator key is essential.

## BASIN WRENCH

To reach tap unions under a basin or bath you will need a basin wrench.

## RADIATOR SPANNER

A radiator spanner is used for fitting radiator tails.

## PUMP NUT SPANNER

Replacing a worn out central heating pump should be easy, but it usually isn't because the nuts are seized up. A tool such as this pump nut spanner gets a much better grip than a normal spanner.

## STEP WRENCH

This is used for jobs such as fitting valves to radiators, connections to tanks and so on.

## BOX IMMERSION SPANNER

A large box spanner is needed to fit or replace an immersion heater.

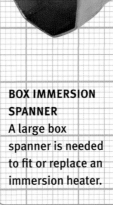

# 2 Pipes, Taps and Valves

Water flows through pipes controlled by valves, and exits via taps. Once you understand these processes, you are well on the way to understanding plumbing.

# Water Supply Pipes

The word 'plumbing' comes from the Latin word for lead, plumbum, because the Romans who developed Europe's first water systems used pipes made from this metal. Some old houses still have lead pipes, but it is expensive and poisonous so has not been used since the early twentieth century. If you discover any lead pipe in your house you need to replace it as soon as possible.

Early central heating systems used cast iron pipe and later steel was used. Copper pipe replaced both. Copper was originally soldered using tin/lead solder, but this has been phased out in favour of lead-free solder. Plastic pipes have been used for waste water for many years, but it is only recently that the development of strong, heat-resistant plastics and simple push-fit fittings has led to its widespread use for water supply.

## Copper Pipes

Copper tube for water supply work is usually 15mm or 22mm diameter (referred to as Ø15 and Ø22) and comes in 2-metre or 3-metre lengths. Like many metals, copper becomes harder when worked and can be softened by heating and cooling. Straight tube is normally supplied 'half-hard', which means that it is tough enough to resist most knocks, but is soft enough to bend. Any tube tends to kink when bent. When bending half-hard copper you must always use either a spring or a pipe bender.

Chrome-plated copper pipe is commonly used in bathrooms, and anywhere that pipework is on show.

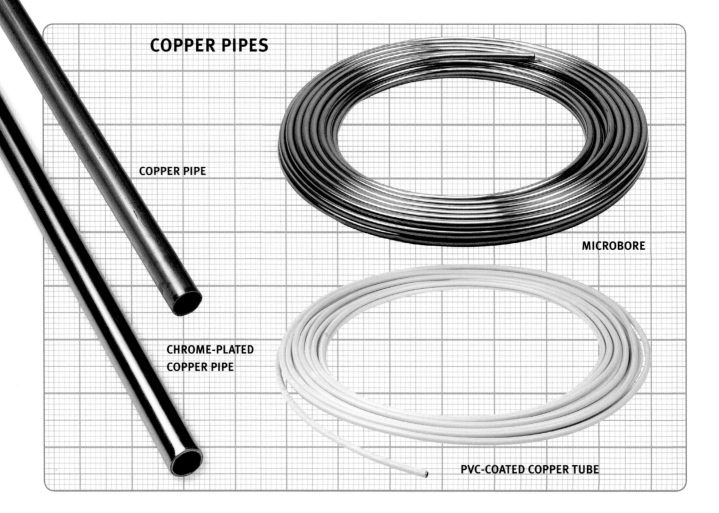

**COPPER PIPES**

COPPER PIPE

MICROBORE

CHROME-PLATED
COPPER PIPE

PVC-COATED COPPER TUBE

## Microbore

Older central heating installations were based mostly on Ø15 copper tube, but since the 1980s it has become more common to use Ø8 and Ø10 tubing, which is known as microbore. This is supplied in a soft state on rolls, which reduces the number of joints. Though care is needed when making tight bends, it can be coaxed into loose curves quite easily.

## Cleaning

Copper corrodes when exposed to the atmosphere, going from a bright red-orange to a dull dark-brown colour. This is not harmful: in fact, the outer layer protects the underlying copper. However, solder will not stick to corroded copper, so the ends of pipes must be cleaned before soldering. Even very new pipes will have some corrosion, not to mention dirt and grease, and need cleaning.

If you use push-fit or compression joint on copper it is essential that the ends of the pipe entering the joint are in good condition. A small dent or even a scratch can make it impossible to get a water-tight seal. Copper tube used with push-fit fittings should be cut very accurately and the end chamfered so it slides into the joint easily without damaging the O-ring.

## PVC-coated Copper Tube

Polished copper pipework looks good in old houses, but most people paint exposed pipework. An alternative is PVC-coated copper tube, which is available in the same range of sizes. The copper is thinner as it is partly replaced by the PVC outer layer. This adds a certain amount of insulation to the pipe, helping to prevent hot water cooling on its journey from boiler to tap or radiator. It also makes the pipes quieter. It is quite expensive and must be ordered specially.

PVC-coated tube is made of soft copper, and even the Ø15 and Ø22 tubes come in long coils, which helps to ensure quicker, neater installations. It cannot be soldered, but either compression or push-fit fittings may be used. As the coating is usually white, white push-fit fittings are an obvious choice.

**DISCREET PIPEWORK**
PVC-coated copper tube is ideal for central heating where it is on display.

**toptip\***

To most people the words 'pipe' and 'tube' are interchangeable. Some people assert that pipes are always round while tubes may be different shapes. Others claim that different manufacturing techniques are used, but catalogues from manufacturers do not bear that out. It's common to see the same product listed as either pipe or tube, and many people buy tube to do pipework. Take your pick!

# Plastic Pipes

Plastic pipes have several advantages over copper: they are lighter and cheaper, and they come in long rolls so there are fewer joints to make and potentially fail.

Plastics can be permeable to oxygen, which encourages corrosion. To prevent this an impermeable barrier membrane is incorporated, which is why such pipes are often described as 'Barrier Pipe'.

Plastic is a better insulator than copper, so hot water stays hot while the outside of the pipe stays cool. Water in plastic pipes is less likely to freeze because of the insulation. If it does, however, plastic will stretch to accommodate the ice, where copper tube might burst or be forced out of the joints. There are several types of plastic plumbing in use.

**PLASTIC PIPE COLOURS**
Don't judge a pipe by its colour – PE-X, PB and CPVC usually come in white (*above*) or grey (*opposite*). Grey is cheaper, and white looks better when the pipework is on show, but the colour does not define which type of plastic it is.

## Anatomy of **Plastic Pipe**

1. Polybutylene
2. EVOH oxygen barrier layer
3. Polybutylene
4. Adhesive ensures secure bonding of polybutylene to barrier layer

## Bend radius for Speedfit PEX

| Pipe diameter (mm) | Clipped (mm) | Cold form bend (mm) |
|---|---|---|
| 10 | 100 | 30 |
| 15 | 175 | 75 |
| 22 | 225 | 110 |
| 28 | 300 | 225 |

## Cross-linked Polyethylene (PE-X)

Cross-linked Polyethylene (PE-X or PEX) is a semi-rigid plastic increasingly used for water supply and central heating work. It comes in the same sizes as copper and can be used with the same fittings. For small jobs 2-metre or 3-metre lengths can be bought, but it is more flexible than copper and even larger sizes are available in rolls.

PE-X can be used with either compression or push-fit joints, but to obtain a good seal the pipe must be in good condition and cut accurately so a pipe cutter is essential. The pipe is softer than copper and needs some support where it enters the joint to prevent it being distorted and causing leaks. The pipe and fitting manufacturers supply inserts to put inside the tube before putting it into the joint. It is essential you use them and make sure they are the right ones for that particular type of joint.

## Polybutylene (PB)

Polybutene-1 is the internationally agreed name for a plastic material approved for use in water supply and central heating work. In the English-speaking world it is commonly known as Polybutylene (or PB).

Polybutylene is softer and more flexible than PE-X, making it ideal for underfloor and in-wall heating, where a lot of pipe has to be laid in coils or zig-zags in a confined area. It can be used for water supply too, but being more flexible you need to use more supports to prevent it sagging. A sagging pipe can pull right out of a joint.

The extra flexibility of PB pipe makes it relatively easy to fit into floor spaces, by threading or cabling it through holes drilled in the joists. Pipes fitted like this are less likely to be damaged by nails than those dropped into slots cut into the top of joists.

Polybutylene pipes are available from Ø10 upwards and use the same push-fit and compression fittings as PE-X. As with PE-X, it is essential that you use the right inserts.

Polybutylene pipe and fittings sold in the UK have to meet BS 7291 and should last for at least 25 years. Since the late 1990s, there have been arguments and legal cases in the US concerning PB installations that failed after relatively short periods of time, which has damaged the reputation of the material there. The situation is complex but many cases are due to poor installation using the wrong fittings, failure to properly support the pipe, and unscrupulous suppliers selling fittings described as PB but actually made from something else. One real concern is that in some states in the US the water supply contains far higher levels of chlorine than is permitted in the UK, which causes degradation of the pipe.

## Chlorinated Polyvinyl Chloride (CPVC)

Chlorinated Polyvinyl Chloride is PVC that has been treated to make it harder and stronger by reacting it with chlorine, and is sometimes known as PVC-C. It is widely used in industrial pipework and for domestic plumbing in the US, but rarely in the UK. CPVC fittings are normally glued.

# Supply Pipe Accessories

## PIPE CLIPS

### NAIL-IN PIPE CLIPS
The simplest pipe clips resemble large cable clips. They are intended for use on pipework that isn't on view – under floors, inside hollow walls and so on. They are small, so ideal when two or more pipes run close together, and quick and easy to fit.

### SCREW-FIXING PIPE CLIPS
Screw-fixing pipe clips are more trouble than nails, but provide a better grip and look neater too. Simple push in types are used where the pipe is not exposed, but pipes on view should be held by clips that lock around the pipe to prevent it being pulled out accidentally.

### TRADITIONAL PIPE CLIP
This alternative to the nail-in clip for copper pipe needs two screws. It is not very common these days. Similarly shaped, but much larger, plastic clips are often used for waste pipes.

### COLD FORMING BEND FIXTURE
Plastic pipe can be bent using springs or pipe benders, but it tends to try to straighten itself, leading to strain on neighbouring joints. Bend fixtures hold the pipe firmly.

# CONDUIT

### FLEXIBLE CONDUIT

Copper pipes can be corroded by the alkali in cement and plaster, so burying them in floors and walls without protection is asking for trouble. Temperature change is also a problem because copper expands more than masonry when heated, leading to movement which puts strain on the pipes and joints. If you need to bury a pipe, the simplest way is to run it through a flexible plastic conduit. Avoid making joints in sections that will be buried; if possible, arrange the pipe run so it can be pulled out and replaced if necessary without digging up the floor.

### PIPE COLLAR

It is impossible to plaster or tile neatly around a pipe. Pipe collars fit neatly around pipes, covering the ragged edges of the hole.

### PIPE COVER

Pipe covers fit over straight runs of pipe. Some require a back plate behind the pipe, but many are designed to click onto pipe clips – provided they are from the same manufacturer.

### RADIATOR BACK BOX

A special form of pipe collar is made for use behind radiators connected to microbore pipes. These are used where the pipes are run down a wall prior to dry lining, then left sticking out of the plasterboard. The outlet covers the rough edge of the board and plaster, and holds the pipe firm preventing damage to the wall when you fit the radiator.

# Water Supply Fittings

## SUPPLY JOINTS

Pipework is nothing without fittings, but before looking at the different ways of attaching them to pipes, it helps if you know their shapes and what they are used for.

### TEE JOINTS
These joints shaped like a letter T are used to join three pipes, making a branch.

### COUPLINGS
Straight joints, normally called couplings (front left), are simply to attach one pipe to another. 'Reducing connectors' (back right) are used to join, for example, a Ø22 pipe to a Ø15pipe.

### ELBOW JOINTS
These are used to connect two straight pipes at a corner without bending the pipe.

### CROSSOVERS
These are preformed to make it easy for one pipe to cross another.

## TANK CONNECTORS

These are similar to tap connectors, but are designed to suit water tanks and hot water cylinders.

## PIPE MANIFOLDS

These are used to connect several pipes. Large ones are used to split central heating pipes down to Ø8 for radiators. Four-way manifolds are useful in bathrooms and kitchens to supply the basin, bath, shower, sink, dishwasher and washing machine.

## FLEXIBLE CONNECTORS

These are used to avoid awkward pipe bending and are invaluable for connecting, for example, a new bath or sink to existing pipework. They may have compression or pushfit fittings on both ends or a tap fitting on one end. They are usually 300mm long but different types are available to suit different pipes and taps.

## TAP CONNECTORS

These are required to match various sizes of pipe with different kinds of taps. Straight and elbow types are usually used under sinks and basins. Wall plate types are generally for outdoor taps and occasionally over 'traditional' sinks.

# MAKE AN END-FEED JOINT

There are several different ways of joining pipes and attaching pipes to fittings, depending on the material used for the pipe, the type of fitting and its location.

End-feed joints are the cheapest and neatest way to join copper pipes. These are made of copper, and the inside diameter of the end-feed joint is slightly larger than the outside of the tube so it can slide in. The joints are held in place by solder, which is fed in from the end, hence the name.

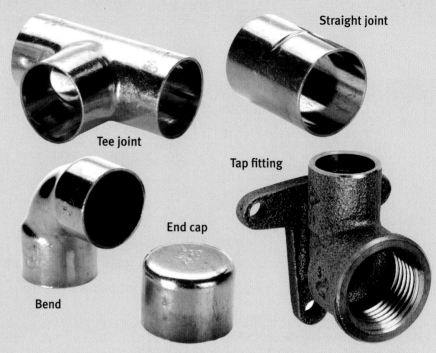

Tee joint

Straight joint

Tap fitting

Bend

End cap

Couplings

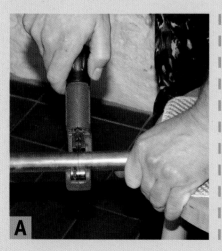

**A**

## top tip*

Soldering takes practice, particularly knowing how long to heat the pipe before applying the solder. Buy several spare joints and an extra piece of pipe to practise on before starting on a real project.

**1** Cut the pipe with a pipe cutter **A**. If you use a saw you will need to file the end so it is smooth and slightly tapered and will slide into the joint. End feed joints have a shoulder in the centre. Both pipes should slide in up to that point.

**2** Pipe cutters leave the outside of the pipe very smooth, but tend to leave a burr on the inside. Designs vary, but most come with a built-in deburring tool to smooth the inside of the pipe **B**.

**B**

### END-FEED FITTINGS

Although end-feed fittings are mostly made from copper, the tap fittings are cast brass.

**3** The ends of the pipes and inside the joint need to be clean or the solder will not flow. To do this, use emery cloth, glass paper, wire wool or specialist cleaning pads/tools **C**.

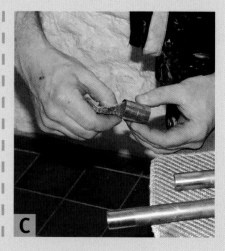

**C**

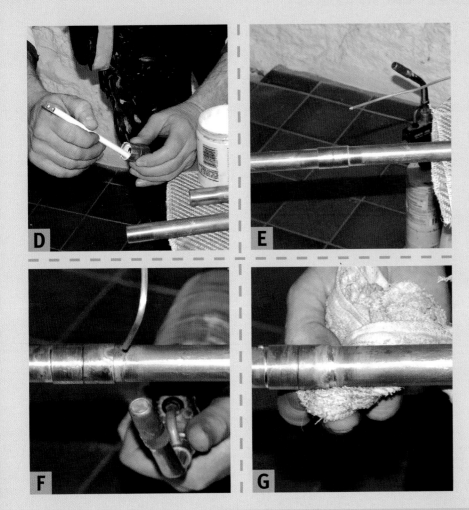

**4** Smear flux inside the joint and round the outside of the pipe **D**.

**5** Put the joint together and wipe off any excess flux **E**.

**6** Heat the joint from underneath, wait a few seconds then apply solder to the top **F**. You must have the pipe hot enough to melt the solder. Do not use the flame to melt the solder directly. If the pipe is hot enough, the solder will melt and be drawn into and right around the joint. If the pipe is vertical, move the flame around it.

**7** Wipe the joint whilst the solder is still molten to remove any unsightly blobs **G**.

## doit MAKE A SOLDER-RING JOINT

Solder-ring joint fittings have an embossed ring round each end that is filled with solder. This is slightly more convenient than using end-feed joints, but is particularly usseful when you are working in an awkward space, as you can make the joint one-handed.

The technique for making solder-ring joints is very similar to end-feed joints. Cut the pipe, clean it and clean inside the joint. Apply flux. Put the joint together. Wipe excess flux off then heat it. When the joint is hot enough a shiny ring of solder will appear around the pipe. Check it goes right round; if not move the flame to encourage it.

Bend

Straight joint

Tee-joint

Reducer

End cap

Reducer

# Compression Joints

Soldered joints are neat, cheap and strong but if you are nervous of soldering, or do not want to buy a blowtorch, flux and solder just for a small job, compression joints are a good solution. They are also useful if moving or adjusting an item might be necessary, and where a naked flame might prove dangerous, such as inside chipboard kitchen units. Conex, the most popular brand in the UK, has become almost synonymous with compression joints, though there are now many others.

Inside a compression joint is a small copper or brass ring called an olive, which slips over the pipe. The body of the joint and the inside of the nut are both tapered so that when the nut is tightened, the olive is squashed on to the pipe. Most plumbers apply a smear of jointing compound to the pipe to improve the seal. Do not use jointing compound on plastic pipes, as it may damage the plastic. If you are using compression joints on plastic, a couple of turns of PTFE tape around the olive is recommended. This is a clean, simple alternative for copper, too.

In order to work, the olive must seal against the pipe and the body of the joint, so it is essential that the pipe is undamaged. If you buy joints from various suppliers, don't mix and match the joints, nuts and olives – slight differences in the design of each can stop them working properly.

Compression joints can be unscrewed and replaced, with a smear of fresh jointing compound or turn of PTFE around the olive, but removing the olive is almost impossible. You can get olive removal tools but it's not something the average DIY person needs to attempt. If you need to move pipes and redesign a system, just cut the olive off and join new pipe as necessary.

# doit USE A COMPRESSION JOINT

**1** Measure and cut the pipe carefully. If you are using plastic pipe, try to to cut on one of the marks (see the example on page 44). If you are using copper pipe, mark the cut position and insertion depth. Use a pipe cutter like this for copper pipe, or a plastic pipe cutter recommended by the pipe manufacturer **A**.

**2** Dismantle the joint and slide a nut and olive on to the pipe. If you are using plastic pipe, fit the manufacturer's recommended insert into the pipe as shown **B**.

**3** Wrap a little PTFE tape around the olive **C**. You might prefer to use jointing compound on copper pipes but it is unsuitable for use on plastics.

**4** Insert the pipe, making sure it goes right into the joint. Tighten the nut by hand **D**.

**5** Tighten it a little more using a spanner or a pair of pipe pliers **E**. DO NOT OVERTIGHTEN – one turn on the nut from 'hand-tight' will usually be sufficient.

## toptip*

A pipe insert must be used when using compression joints on plastic pipe, and it must be of the type that is recommended by the manufacturers. Some manufacturers provide different inserts for use with compression or push-fit joints.

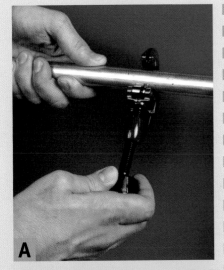

**A**

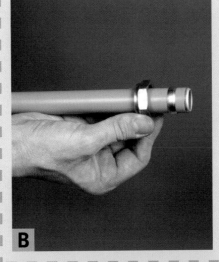

**B**

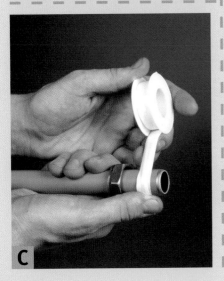

**C**

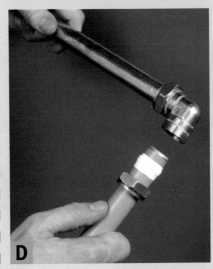

**D**

**E**

Pipes, Taps and Valves **43**

# Push-fit Fittings

Recent advances in plastics technology are making plumbing easier. There are now long lengths of pipe that will reduce joints to a minimum, as well as joints that simply push on to pipes and lock in place.

The illustration below shows a push-fit joint. The actual seal is made by an O-ring, the circular seal that presses tight against the body of the joint and around the pipe, but that would be no use without a means of holding the pipe in place against the force of the water. This is done by a collet, a ring of plastic that fits around the pipe and has 'fingers' on the inside that are fitted with sharp stainless steel teeth. The collet is designed so that under pressure, the pipe moves out of the joint a little and the teeth dig in tighter.

You can release the joint by pushing the collet back into the joint body and holding it there as you pull the pipe out. To stop this happening accidentally, you can get clips that fit around the collet preventing it from moving. Some are even colour-coded to distinguish hot and cold pipes.

Some ranges of push-fit fittings use a twist-to-lock system, which gives added security. The sealing system is the same as any other, but a half-turn on the screw cap tightens the collet and prevents the pipe being pulled out, so there is no need to use clips on the collet.

Other ranges have a metal shroud over the collet so it cannot be moved accidentally. The only way to release this is by using the special release tool.

## toptip*

**It is important that you use the correct insert for the joint. Do not mix 'n' match manufacturers.**

## Anatomy of a **Push Fit Joint**

1 Grips before it seals
2 Stainless steel teeth grip the pipe
3 Pipe insert gives secondary seal
4 Main 'O'-ring seal

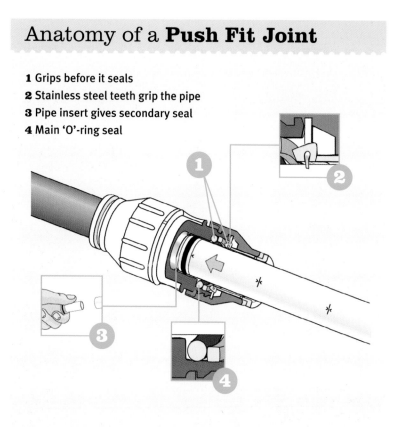

## CUPROFIT FITTINGS

These copper fittings are very strong, but are as easy to use as the plastic push-fit fittings.

CUPROFIT
TEE

CUPROFIT
TAP FITTING

# USE A PLASTIC PUSH-FIT JOINT

Push-fit fittings can be used on plastic or copper pipe (though not chrome-plated copper sometimes used in bathrooms because the chrome layer is too hard for the teeth to get a grip).

**1** If you are using plastic pipe, cut it on one of the marks shown by the ✂ symbol. When you insert the pipe into the joint, it will go in as far as the next mark. If the installation requires you to cut it off at a mark, treat it as copper pipe as in 3 below **A**.

**2** Put the supporting insert in the end of the pipe **B**.

**3** If you are using copper pipe (or plastic not cut on the mark), hold the joint next to the pipe as shown and mark the insertion depth **C**.

**4** Push the pipe into the joint up to the mark **D**. If you are using Hep2O or Speedfit, tighten the locking ring. If you are using FloFit, fit a collar. Even when tightened the fitting can be rotated to align with other pipes **E**.

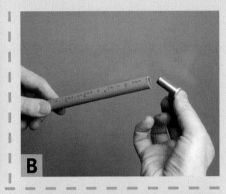

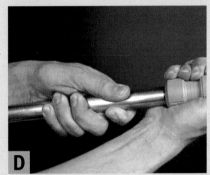

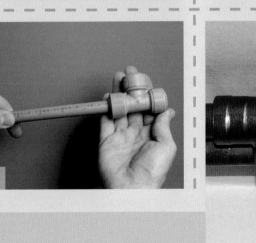

## Conex Cuprofit

In situations where pipework might be subjected to rough treatment, such as garages and workshops, copper pipe is more durable. It makes sense to use metal fittings as well. Until recently, that meant soldering or compression joints, but now there is an alternative. Cuprofit joints use O-rings and collets just like the plastic versions, but they are made from copper and brass. The original Cuprofit joints may be used with any plastic or copper pipe, but are complex, bulky and expensive.

In DIY stores Cuprofit is marketed under the Masterflow brand. Conex's Cuprofit 2 range is much slimmer, simpler and cheaper, but can only be used on copper. At the time of writing however, these were available in only a limited range of shapes and sizes.

**METAL ON METAL**
Metal fittings are ideal for pipes that may suffer rough treatment (*right*).

# Taps and Valves

**TAP REVIVER**
Old-fashioned or worn-out taps can be renewed easily with a kit such as this (*right*).

**TAP RESEATING TOOL**

Taps and valves are used to control the flow of water. The only difference is that valves control water passing from one pipe to another, or into or out of a tank, boiler or radiator, while taps let water run out into sinks, basins and baths.

## Traditional Taps

Until recently, most taps and valves operated in a similar way: turning a handle operated a screw that raised or lowered a seal, which pressed on to a 'seat'. Replacing tap washers is less common now, but a dripping, old-style tap it is not very difficult to fix (see *Fix a Dripping Tap*, pages 70–71).

## Tap Revivers

If you have an old-style tap that you wish was a bit more modern, a 'tap reviver' such as the model shown, may be suitable. The old tap is dismantled then, rather than changing the washer, the new cartridge is simply screwed in and the new top fitted.

## Tap Reseating Tool

Sometimes the seat in a tap becomes worn and pitted, and the washer does not make a good seal, even when new. If you are really determined to keep the tap (perhaps because it suits your décor), you will need either a tap reseating tool to flatten the seat, or a reseating kit, which should include plastic seat inserts as well as cones to replace the conventional flat washer.

## Ceramic Disc Taps

Modern taps rely on ceramic discs rather than rubber washers to make a seal. There are probably thousands of designs but they all operate in the same way. The disc has a hole (or slot) in one side through which water can pass when the hole lines up with a matching hole in the tap body. Rotate the disc a quarter turn and the flow stops.

Ceramic disc taps have many advantages: ceramic is a lot harder than rubber so the discs last a long time; they are easy to operate – many have a lever you can operate with an elbow when your hands are full or filthy; the need for only a quarter turn gives designers more scope for making attractive shapes. Nothing is perfect though, and discs do eventually wear and start to leak. When this happens, you will need to replace the cartridge that houses the disc and the mechanism controlling it.

## Types of Tap

Despite the thousands of designs, there are actually only three basic types of tap:
- single taps, one hot and one cold
- deck mixers with a wide body that cover two holes in the sink/basin where the hot and cold water connections are made
- monobloc mixers that fit in a single hole in the sink/basin.

Strictly speaking, mixer taps for kitchens are not really mixers at all. Allowing hot and cold water to mix in the tap may in some circumstances lead to hot water from a house entering the cold main and, as many hot water systems are fed from a tank in the loft, this could lead to contamination of the supply. Kitchen 'mixers' have concentric outlets: the hot water comes through a hole in the middle of the spout and the cold is sprayed round the edge so the water mixes in the air, not in the tap.

Until recently, mixer taps in bathrooms were fed by cold water coming from a tank in the loft, so there was no danger of backflow into the supply system. However, modern systems generally dispense with the tank and run both hot and cold at mains pressure, so check valves are required on mixer taps and showers.

Monobloc taps are designed to fit in a single hole in the sink/basin. It's neat and convenient, especially for twin sinks, where a monobloc can fit between them and serve both, but there simply isn't room for conventional tap fittings. Monobloc taps are fitted with a couple of 'tails' – pieces of soft copper pipe or flexible connectors about 300mm long that can be joined to the water supply pipes (see pages 70–73 for advice on repairing or replacing taps).

**CERAMIC DISK VALVE**

# Valves

## Saddle Valves

Saddle valves sit on a pipe like a saddle on a horse, and have a section that goes round the pipe totally enclosing it. Unlike a saddle though they have a sharp hollow point underneath so that when the valve is clamped securely on to the pipe the point penetrates, allowing water to flow out of the valve. Saddle valves are included in garden tap kits and plumbing kits for fridges with built-in ice makers. (See *Fit a Garden Tap*, pages 74–75).

Some saddle valves have a compression fitting and some a push fit, but kits that are sold for this purpose often have a valve with a screw fitting like those on washing machine valves (see page 50) so a flexible hose can be used to connect to the tap.

## Stopcocks

'Cock' is an old, but still common English term for a valve. The stopcock is the valve that cuts off the water supply to your house. Since water is usually piped underground, the stopcock will be close to the floor where the rising main enters your property, very often in the kitchen under the sink, but maybe in a garage, utility room or anywhere else the builders found convenient. Since stopcocks are rarely turned off, they rarely wear out, but they do suffer from corrosion and can become very hard to turn. It's a good idea to find it and test it occasionally to make sure you can turn off the supply. If you have a leak and need to do this in a hurry you'll be glad you did. You need to turn off the stopcock before carrying out many of the supply jobs in this book.

**GOOD CONNECTOR**
A saddle valve is a quick and easy means of connecting an outdoor tap, ice-maker or washing machine (*above left*).

**BOUNDARY**
The company stopcock is usually found underground (*above*).

The company stopcock is usually situated on the boundary of your property. It seldom needs to be touched unless the domestic stopcock or the pipe that leads to your house is damaged. It is increasingly common to find a water meter beside the company stopcock, with a wad of insulation in the chamber to prevent the meter and pipework freezing up in bad weather. Be sure to replace this.

If you are unable to turn off the company stopcock you should contact your water company. Everything in the water supply system up to and including that external stopcock is its responsibility.

Once the supply has been isolated and the system drained (see *drain cock,* right), a domestic stopcock can be dismantled in exactly the same way as a tap. Most are still made of brass with a traditional tap washer, but you may find more modern ones in new properties. If it is old, corroded and seized up, it is probably easier to replace it.

## Drain Cocks

Drain cocks enable you to drain water out of system you want to work on. There should be one above the stopcock to empty the supply pipes (see photograsph above left). If your hot-water system is fed by a tank in the loft, there will be another drain cock on that system and another on the central heating system (see *Radiator Valves,* page 53). Drain cocks may have a spigot to which to attach a hosepipe, or a permanently-fitted drain that connects into the waste system or simply passes out through a wall to a gulley.

If you use a hose connected to a drain cock, more water will drain from from the system if you pass it out though a door or window and hold the open end lower than the drain cock. Once the flow is established, most of the water will syphon out – even over a windowsill – leaving only a small amount left in the hose. If you run the hose into a sink, then all the water in the system that is lower than the sink will remain there.

**IN POSITION**
A drain cock next to the stopcock allows you to drain the system (*above left*). The heating drain cock (*above*) is fitted on to the lowest point of a central heating system.

## Service/Isolation Valves

Whether you call them service or isolation valves, these valves are fitted close to appliances so that you can isolate them while you work on, or service them. In modern buildings you will find an isolation valve on the feeds to every tap, WC cistern, shower unit and boiler, but in older properties they are rare.

Isolation valves may resemble old-style stopcocks, but modern ones are much smaller. They have a ball inside with a hole passing through it. The ball can be turned either with a handle or a screwdriver so that the hole is either inline with the body of the valve allowing water to flow, or across it turning the water off. Isolation valves must be fitted the right way round since they rely on the water pressure to press the ball against an internal seal. Look for an arrow on the valve body to show which way the water should flow.

If you are working in a property without isolation valves you will have to switch off the stopcock and possibly the drain header tanks before you can replace a tap washer or a WC syphon. This takes time and prevents anyone using the rest of the plumbing in the house, so whilst the water is off, it makes sense to fit isolation valves, at least in the area where you are working.

## Washing Machine Valves

Washing machine valves are very similar to isolation valves. The only difference is that while an isolation valve has compression or push-fit fittings on both ends, the outlet end of a washing machine valve has a screw fitting to suit a washing machine hose. Exactly the same fitting can be used for dishwashers, fridges with ice-makers, and even outside taps connected by a hose.

## Check Valves

Check valves allow water to pass in one direction but not the other. They are used where there is a danger that a drop in mains pressure could draw potentially dirty water back into the supply. They are used for mixer taps and showers connected directly to the supply or outside taps that could suck in water from a hosepipe. Double-check valves are simply two valves in the same body – if one fails the other should still prevent a backflow.

Check valves are fitted just like isolation valves. Some outside taps or bath/shower mixers have them built in.

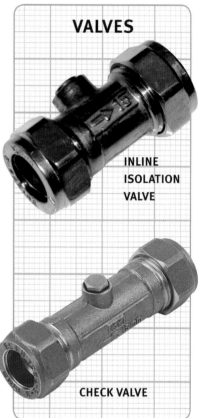

**VALVES**

INLINE ISOLATION VALVE

CHECK VALVE

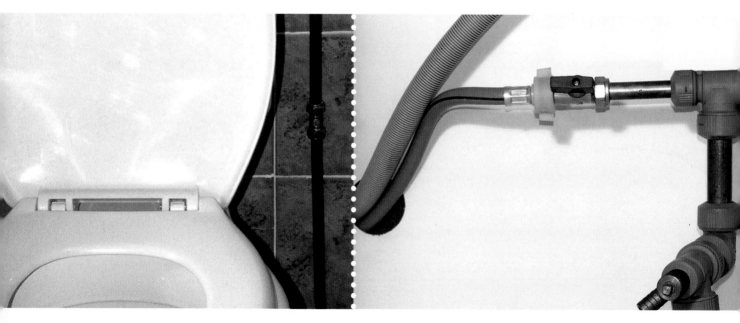

**CHECK VALVES**
When mixer taps are connected directly to the supply, a drop in pressure may draw potentially dirty water back into the supply – but check valves cut out this risk (*right*).

---

**APPLIANCE VALVE**
Washing machines and dishwashers use isolation valves with colour-coded handles (*left*).

---

**ISOLATING AN AREA**
An isolation valve allows you to work on the WC without turning off the main stopcock (*far left*).

---

### RETRO STYLE
Reproduction radiators in many styles are available to complement the period of your house (*above left*).

### ULTRA MODERN
Up-to-the-minute designs are best for modern homes or apartments (*above right*).

### ANTIQUE RADIATOR
Treasures found in reclamation yards may be reconditioned (*right*).

## RADIATOR VALVES

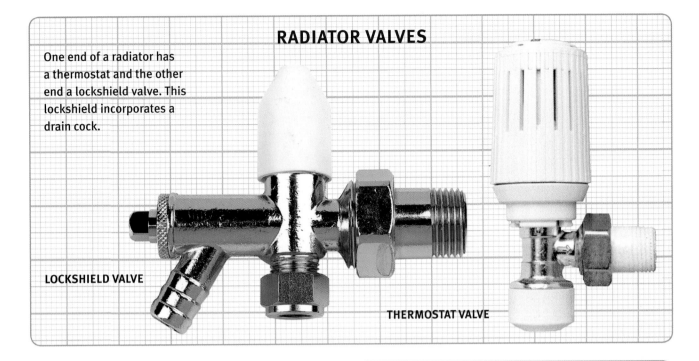

One end of a radiator has a thermostat and the other end a lockshield valve. This lockshield incorporates a drain cock.

**LOCKSHIELD VALVE**

**THERMOSTAT VALVE**

## Radiator Valves

Radiators have valves on both ends. One will have a control knob, possibly with a built-in thermostat, the other end just has a cover over the spindle. This second one is called a lockshield valve. It is identical to a non-thermostatic valve without the knob.

Lockshield valves serve two purposes: by turning it off along with the valve on the other end you can remove and replace a radiator that has sprung a leak, or just remove it whilst you decorate. The valve shown above incorporates a drain valve and should be fitted to the lowest radiator in the house to facilitate draining.

The lockshield can also be used to 'balance' the radiator output. The heat output of a radiator is rarely ideal for a particular room. Part-closing a lockshield valve on a radiator that is giving out too much heat reduces the output so that more is available to radiators elsewhere in the house. With thermostatic valves more common, this is less common than it once was.

When a heating system is first filled, the air inside it has to escape to let the water in. Bleed valves are fitted to the top of each radiator for this purpose (see *Bleed a Radiator*, page 109).

New radiators are supplied with threaded holes in each corner. You need to fit radiator tails to the two bottom ones, a bleed valve to one of the top ones, and a blanking plug to the remaining one. Double radiators may vary, and some need two bleed valves.

There are different types of radiator tails to suit different valves, so check when buying them.

## GATE VALVE

Gate valves intended for domestic use are operated by a turnwheel on a screw mechanism. Inside is a sliding part – the gate – with a hole in it. When the hole lines up with matching holes either side of the valve body, water flows through. They are used on central heating systems because the path through an open gate valve is smooth and straight so the water flows easily, unlike valves that use washers or disks. This is important as the water in a central heating system is moved around by a low-powered circulator, so anything that impedes the flow will reduce the efficiency of the system.

**GATE VALVE**

# Waste Pipes

**INSIDE STACK**
Modern houses like this have all the waste pipes hidden inside.

Although you might find metal waste pipes in old houses, plastic has been in common use since the 1960s. Waste pipes don't operate under the same sort of pressure as supply pipes so they are larger to allow the waste water to drain away easily. Since they don't need to contain mains pressure, waste pipes are made of thinner material. Waste pipes should slope down towards the drain to encourage the flow and help prevent blockages. To help you achieve this, you can buy elbows and tee fittings made to 92.5° rather than 90°.

Four common sizes are available for different applications, though these are not always adhered to. Plastic waste pipes were introduced to the UK prior to metrication so odd sizes like 43mm are also to be found.

## Waste Pipe diameters

| | |
|---|---|
| Overflows | 21mm |
| Hand Basins | 32mm |
| Baths and Sinks | 40mm |
| Soil Pipes | 110mm |

Several different types of plastic are in use:

## Acrilonytrile Butadiene Styrene (ABS)

ABS is a strong thermoplastic widely used not just for plumbing but for such disparate items as car parts, musical instruments and golf clubs. It can made in any colour, but waste pipes are limited to grey and white. The grey is a little cheaper and perfect for pipes that are out of sight. ABS pipes can be joined by solvent welding, compression or by push-fit fittings. They degrade when exposed to sunlight, so waste pipes that run outside should be painted.

## Polyvinyl Chloride (PVC)

PVC comes in many varieties. MUPVC (or MuPVC), which is sometimes used for waste pipes has recently been redefined by two European standards: PVC-U and PVC-C, though not all suppliers have changed their catalogues yet. PVC pipe is stronger than ABS and more heat-resistant, so it is preferred in commercial premises. It is also more resistant to ultraviolet light, so need not be painted when used outdoors. It is, however, more expensive. Alternatively, use UPVC. PVC and ABS pipes are made in the same sizes, fit into the same fittings, and the same solvent weld adhesive may be used on both. ABS fittings may even be used on PVC pipe or vice versa.

## Polypropylene

Both Hepworth and Marley supply waste pipes made from polypropylene, but their systems are made to a different standard from the common ABS and PVC ranges so pipes and fittings are not interchangeable. You cannot solvent weld polypropylene, but the push-fit system allows for quick installation.

## PLASTIC PIPES

**POLYPROPYLENE**
Often used in commercial premises, polypropylene waste pipes are rarely used in homes.

**POLYVINYL CHLORIDE**
PVC is interchangeable with ABS for waste pipes; UPVC is preferable outside.

**ACRILONYTRILE BUTADIENE STYRENE (ABS)**
White or grey ABS is a common material for waste pipes.

# Waste Fittings

## Types of Fitting

As you might expect, waste pipe fittings come in a similar selection of shapes as supply joints:

## Expansion Coupling

If you are using push-fit waste fittings, leave each cut pipe just a little short (6–10mm) so that it has room to expand into the joint when carrying hot water. Obviously solvent weld joints will not move, so expansion couplings are sometimes necessary on long straight runs. These have a solvent weld joint at one end and a push-fit style fitting at the other.

Expansion couplings are rarely necessary in houses where the pipe runs to the stack are usually short, but might be if you fit, for example, an en suite bathroom in the front bedroom and run a pipe right through to the back of the house.

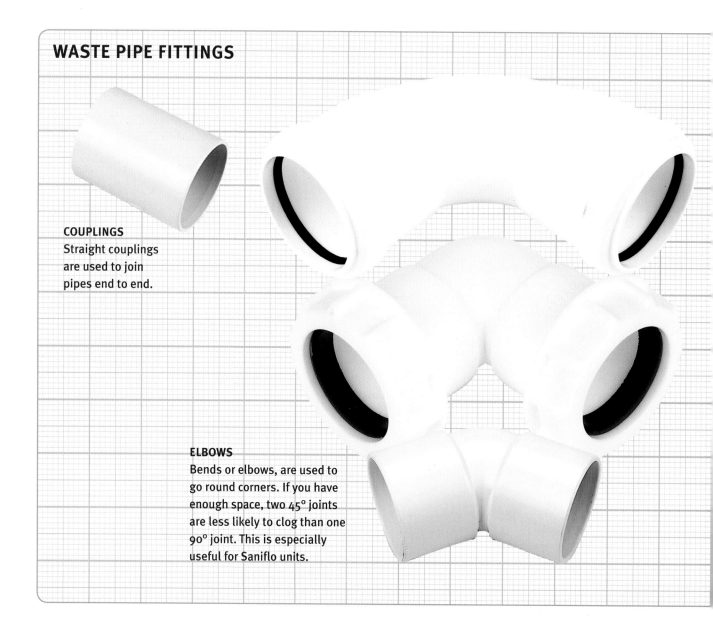

## WASTE PIPE FITTINGS

**COUPLINGS**
Straight couplings are used to join pipes end to end.

**ELBOWS**
Bends or elbows, are used to go round corners. If you have enough space, two 45° joints are less likely to clog than one 90° joint. This is especially useful for Saniflo units.

# Anatomy of an **Expansion Coupling**

An expansion coupling allows long runs of waste pipe to expand without buckling when hot water flows through.

**1** Waste pipe
**2** Expansion coupling
**3** Pipe clip

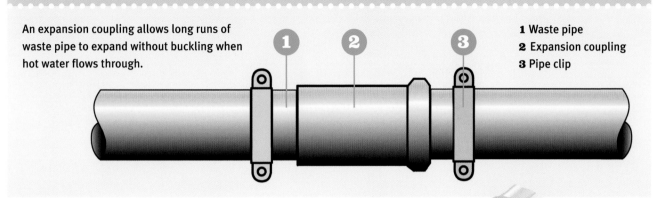

## ADJUSTABLE BENDS

These are useful when carrying out alterations. They allow you to connect, for example, a new bath to the old waste pipe. Adjustable joints are in three sections which make a 90° bend when lined up, but by rotating the sections all sorts of weird and wonderful shapes are possible. If this does not offer enough flexibility, short, flexible waste pipes might be the answer. Adjustable and flexible joints are only available with compression fittings because their flexibility would allow them to slide out of push-fit joints too easily.

**FLEXIBLE WASTE CONNECTOR**

**ADJUSTABLE BEND**

**TEE JOINT**

## TEE JOINT

These are usually 'swept' – that is, the incoming pipe is curved to direct the flow into the straight-through pipe, so it's important to fit them the right way round. They are generally not quite at a right angle either so that the incoming pipe can have a slight slope to it, but 90° joints are available for joining pipes at room corners and in similar situations.

# SOLVENT-WELD A JOINT

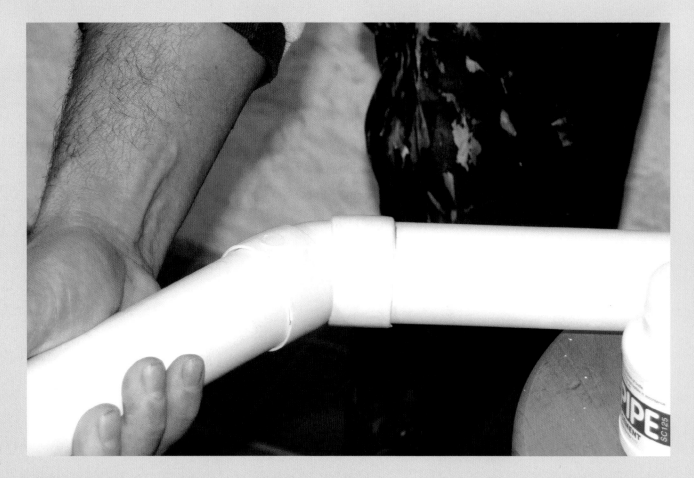

**STRONG JOINTS**
Solvent welding is the strongest, neatest and cheapest way to join waste pipes (*above*).

There are as many different ways of joining waste pipes as there are supply pipes, and most are similar. Solvent welding is the cheapest method and results in the strongest, neatest system, so it's the normal method for professionals. In many respects it is similar to soldering copper pipes. The adhesive contains a solvent that dissolves the surface of the pipe and fitting. The plastics merge, and when the solvent evaporates, they stick tight. If you saw through a solvent-welded joint you will not be able to see the join.

There are disadvantages to solvent welding: you have to get it right first time, though you can do a dry-run to test all the pipes and joints fit BEFORE you glue them; you cannot easily dismantle the system to deal with blockages, and you need to buy adhesive that will mostly be wasted if you are only doing a small job.

1 Cut the pipe taking care to cut it 'square'. A hacksaw will do, but it tends to wander. A 'pull saw' like this will cut much straighter **A**.

2 Remove rough edges from inside or outside the pipe with an emery cloth or small file **B**.

3 Push the joint on to the pipe as far as it will go and mark the pipe with a pencil **C**. Then, when you glue the pipe you will know it is pushed fully home.

4 Remove the joint from the pipe and brush solvent weld cement inside the joint AND around the pipe **D**.

5 Push the joint on to the pipe, checking that it comes right up to the pencil mark **E**. To ensure the cement is well spread, rotate the pipe or joint a quarter of a turn.

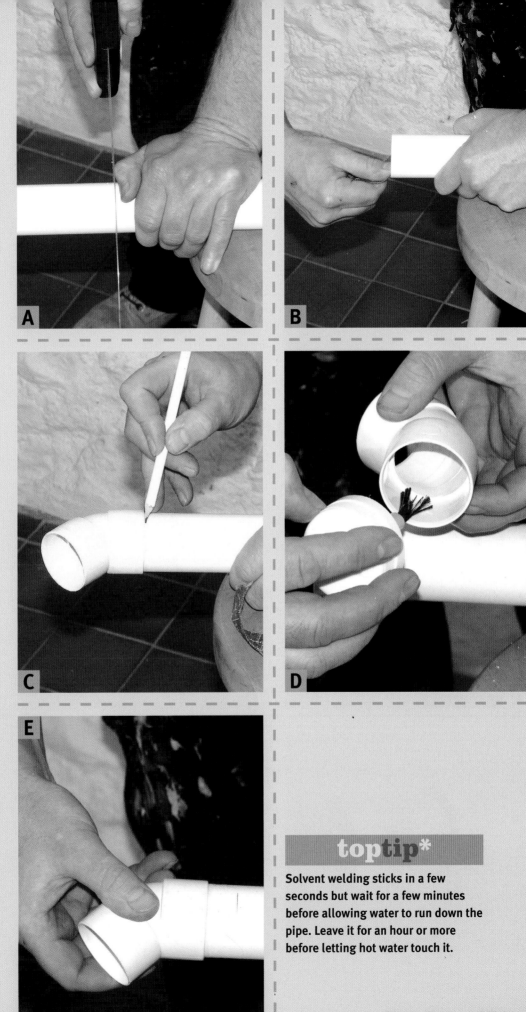

**A**

**B**

**C**

**D**

**E**

### toptip*

Solvent welding sticks in a few
seconds but wait for a few minutes
before allowing water to run down the
pipe. Leave it for an hour or more
before letting hot water touch it.

**Pipes, Taps and Valves**

A

B

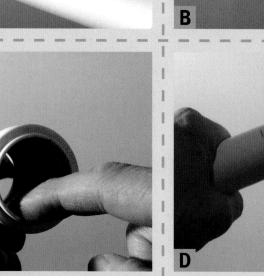

C

D

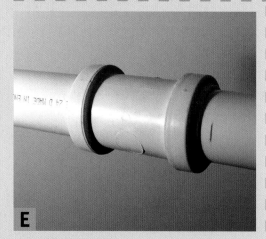

E

## toptip*

Check the diameter of the waste pipe. Push-fit fittings have to be the right size – too small they don't fit, too big they leak. A pipe that looks Ø40 may well be Ø42 or Ø43, so measure carefully BEFORE you go shopping!

Push-fit joints for waste pipes do not have to withstand the pressures faced by water supply pipes, so they are just a close-fitting socket with an O-ring inside to seal round the pipe.

**1** Cut the pipe as for solvent welding (see page 58), then chamfer the edge with a small file **A**. A rough edge to the pipe may damage the O-ring and prevent it making a seal.

**2** There is usually some form of line or 'pip' to indicate the middle of the joint. You can use this to estimate how far the pipe should go into the joint. Mark it with a pencil **B**.

**3** Some makers recommend silicone grease on their O-rings. If you don't have any, some soap on a wet finger will do, or liquid hand wash is even easier **C**.

**4** Push and twist the joint on to the pipe **D** and check that the pencil mark is about 6–10mm from the joint **E**. This is to give the pipe room to expand when hot water runs through it.

# WASTE PIPE FITTINGS

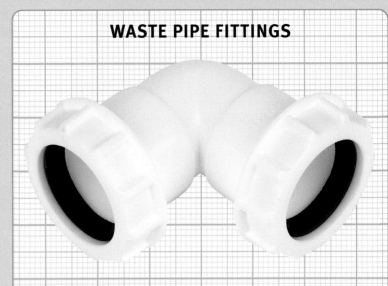

### UNIVERSAL COMPRESSION

Compression fittings are not normally necessary for waste pipes, but 'uni-compression joints' are more tolerant of varying pipe sizes. They are useful if you want to connect to imperial (inch) sized pipes, or join plastic to metal, or polypropylene to ABS. Fit uni-compression joints in the same way as push-fit joints (see page 45), then tighten the nut by hand. You should not need to use a wrench on plastic joints.

### FLEXI WASTE FITTINGS

Compression systems operate by squeezing an O-ring on to the pipe and have a limited tolerance on the pipe size. Flexi Waste fittings from Screwfix consist of a short length of synthetic rubber hose (like a car radiator hose) and a couple of jubilee clips, and will fit a wider range of sizes.

# Waste Traps

ome fairly unpleasant stuff goes down waste pipes, so it's hardly surprising that they smell. To avoid drain odour pervading the house you must have 'traps'. The principle of a trap is quite simple: the waste pipe forms a U-shape, so that some water remains in the bend, preventing foul air from coming back up the waste pipe.

There are different designs of waste trap for different situations:

## S Trap

Originally S traps were hand-made by plumbers bending pipe, and the earliest ready-made traps just copied the shape. These days they are usually made in two plastic sections shaped like a U and an inverted U so that when joined they look like an S on its side. S traps are used on sinks and washbasins when the waste pipe passes down through the floor. Being in two sections makes it possible to configure the joint to line up with the outlet and the waste pipe, and it also makes them easy to take apart and clean if they get blocked.

## TYPES OF WASTE TRAPS

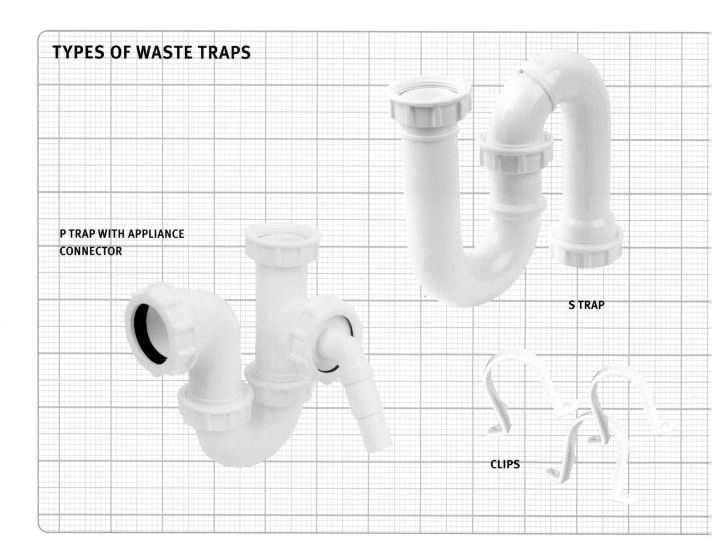

P TRAP WITH APPLIANCE CONNECTOR

S TRAP

CLIPS

## Washing Machine Trap

Washing machines normally empty into a 'standpipe'. These too need a trap, so kits like the one shown here are available. An alternative is an appliance trap such as the P trap shown, bottom left, but some machine manufacturers warn against them as it might be possible for waste from the sink to get into the machine.

## P Trap

A P trap is a variation on the S trap, designed to connect to a waste pipe that passes out through the wall. It resembles a letter P on its side. The version shown here is an 'appliance trap' with a connection for a dishwasher or washing machine outlet.

S and P traps are preferable to other designs for use in kitchens since they are less likely to clog with grease and scraps of food.

## Bath Trap

The bath trap is a low profile version of an P trap moulded in one piece. These are used under baths where there isn't usually room for a full S trap. The version shown here has a connection for a flexible pipe from the overflow. The alternative is a 'banjo' that fits around the waste outlet above the trap, but these make the whole assembly taller.

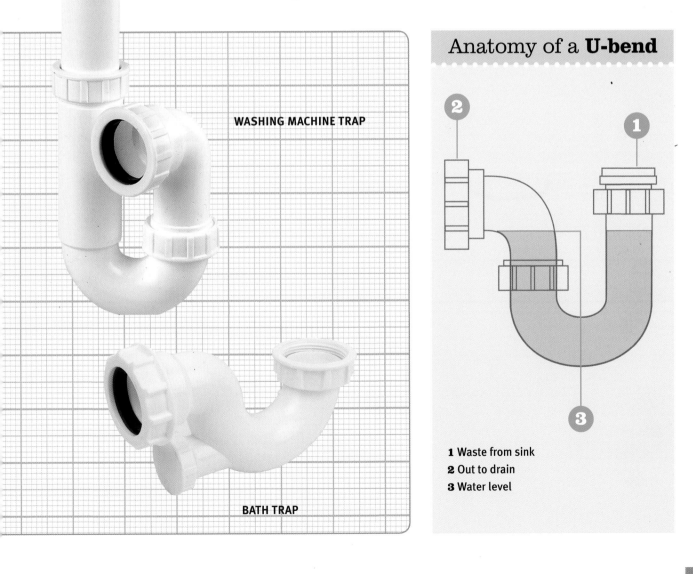

WASHING MACHINE TRAP

BATH TRAP

## Anatomy of a **U-bend**

1 Waste from sink
2 Out to drain
3 Water level

## Bottle Traps

S and P traps take up quite a lot of space, which isn't a problem under kitchen sinks but they look unsightly if used with pedestal washbasins. The bottle trap is slimmer and designed to be hidden by a pedestal. The 'pedestal trap' is an even slimmer variant with the outlet at the bottom.

## Chrome-plated Bottle Trap

When a hand basin is wall mounted without a pedestal to hide the plumbing, a chrome-plated bottle trap and waste pipe looks a lot better than white plastic.

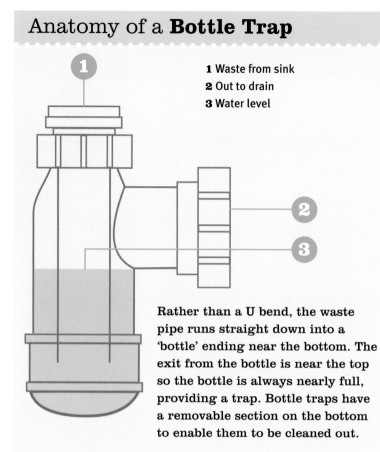

## Anatomy of a **Bottle Trap**

**1** Waste from sink
**2** Out to drain
**3** Water level

Rather than a U bend, the waste pipe runs straight down into a 'bottle' ending near the bottom. The exit from the bottle is near the top so the bottle is always nearly full, providing a trap. Bottle traps have a removable section on the bottom to enable them to be cleaned out.

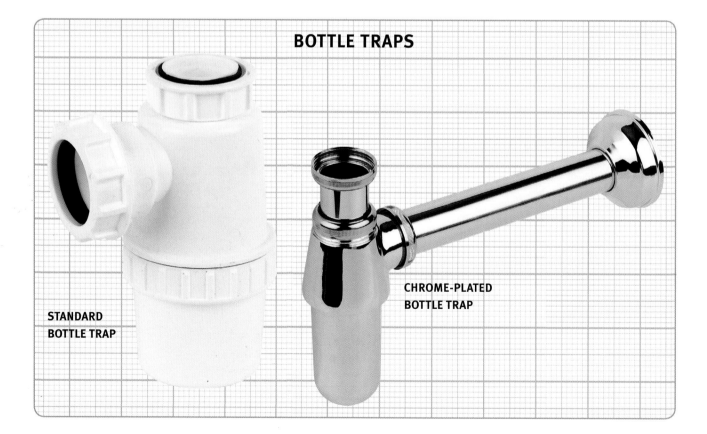

**BOTTLE TRAPS**

STANDARD
BOTTLE TRAP

CHROME-PLATED
BOTTLE TRAP

## SHOWER TRAP

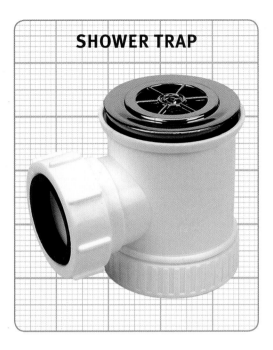

## Anti-Vacuum or Anti-Syphon Traps

Where two or more devices are connected to a single waste pipe, water draining from one of them can lower the pressure in the pipe leading from the other, which results in a gurgling noise from the trap. In extreme cases, such as a bath and washbasin connected to a long run of shared pipe, emptying the bath can suck so much water out of the basin trap that it no longer provides a smell barrier.

This shouldn't happen in well-designed systems, but sometimes compromises are necessary, especially when refurbishing older property, so the simple answer is an Anti-Vacuum or Anti-Syphon trap. These look similar to conventional bottle traps but have a cunning reservoir inside that cannot be syphoned.

## Shower Traps

Shower traps are a special type of bottle trap. People wash their hair in showers and shed a lot in the process. The relatively small amount of water used is not enough to wash away all the hair, so inevitably the shower trap gets blocked. As showers are built in, getting underneath to remove a trap isn't really an option, so shower traps are designed to be cleaned out from the top.

## do it FIT A TRAP

When it comes to fitting, traps fall into three types: most of them connect to an outlet on a sink, basin, bath or bidet. Shower traps are combined with the outlet, and washing machine stand pipes are just open to the air. All types, of course, have to connect to a waste pipe and this is done with a form of compression joint.

The first task is to make sure you have the right trap to suit the outlet and fit your waste pipe – simple if the original installer used standard pipe. Check carefully and if necessary get some reducing couplings or flexible joints before you pull the old pipes apart.

1 If you are installing a new sink/basin/bath or bidet, fit the outlet before you fix the appliance in place **A**. Fit the taps too; it's much easier than kneeling with your head wedged underneath it later. If the waste outlet on an existing appliance looks a bit stained or corroded, replacing it will only add a few minutes and a little extra cost to the job. Outlets are simply fitted with a rubber washer between the flange and the appliance, and another between the underside of the appliance and the nut. The rubber will make a good seal without any heroics. If you over-tighten it you may squeeze the rubber washer out of the joint, and on plastic outlets in particular you may damage the screw thread.

2 The trap screws on to the bottom of the outlet and has a rubber O-ring or washer inside to seal against the end of the pipe **B**. If you have to remove a trap to unblock it, inspect this seal before you screw it back on because it may need replacing.

3 Shower traps are easier still. The upper part screws into the lower with rubber washers either side of the shower base. It is easier to hold the upper part and rotate the lower one.

A

B

# Drains

aste pipes have to empty somewhere and there are two options: a drain (gully) or the soil stack. In most cases, both drains and stacks connect to a public sewer usually running underneath the road.

## Gullies

Older houses often have kitchen sinks running into gullies or drains outside the kitchen. This is legal but they can get smelly in the summer, which is not ideal, especially if you have the kitchen window open. Running individual pipes from baths, showers and basins upstairs down to a drain isn't really an option, and combining them leads to problems of syphoning or backflow. Consequently, it is standard practice these days for all waste pipes to lead into a single soil stack.

**MOVING ON**
Old gullies such as the one pictured above are rare these days. Modern stack systems are much cleaner.

## Anatomy of a **Soil Stack**

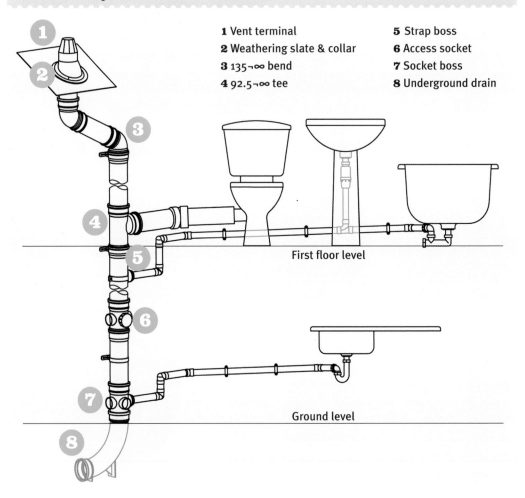

1 Vent terminal
2 Weathering slate & collar
3 135¬∞ bend
4 92.5¬∞ tee

5 Strap boss
6 Access socket
7 Socket boss
8 Underground drain

First floor level

Ground level

## Soil Stacks

Soil stacks are large waste pipes running direct into the underground drain. They are usually Ø110 plastic, although Ø160 is used in large buildings. Older houses usually have the soil stack running down the outside of the house generally close to the WC to make it easy to connect the large WC outlet to it.

In modern houses it is common for the stack to be inside, against an outside wall, so that the drain does not need to run under the floor, but somewhere near the middle of the house so that various downstairs cloakrooms, en suite showers as well as the traditional kitchen and bathroom can all connect to it.

Soil stacks are normally open to the atmosphere to prevent problems with pressure changes when waste flows into them. Traditionally they are made taller than the house and have a vent on the top to keep out debris and birds, but in some cases a stub stack can be used.

A stub stack is just a little taller than the highest waste outlet leading into it, and is capped off with a screwed lid that can be removed to clear blockages. It can only be used if the highest outlet is less than 2m (2.5m in Scotland) above the bottom of the underground drain (known as the invert). This makes it useful for ground-floor cloakrooms and single-storey extensions.

## Anatomy of a **Stub Stack**

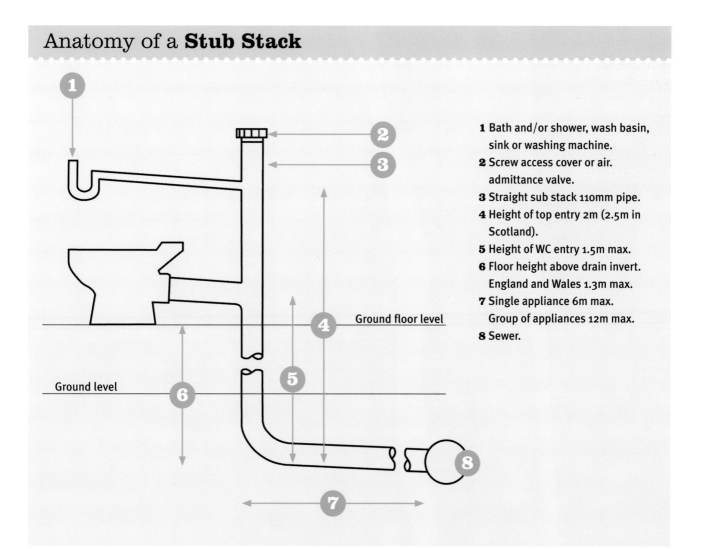

1 Bath and/or shower, wash basin, sink or washing machine.
2 Screw access cover or air. admittance valve.
3 Straight sub stack 110mm pipe.
4 Height of top entry 2m (2.5m in Scotland).
5 Height of WC entry 1.5m max.
6 Floor height above drain invert. England and Wales 1.3m max.
7 Single appliance 6m max. Group of appliances 12m max.
8 Sewer.

Ground floor level

Ground level

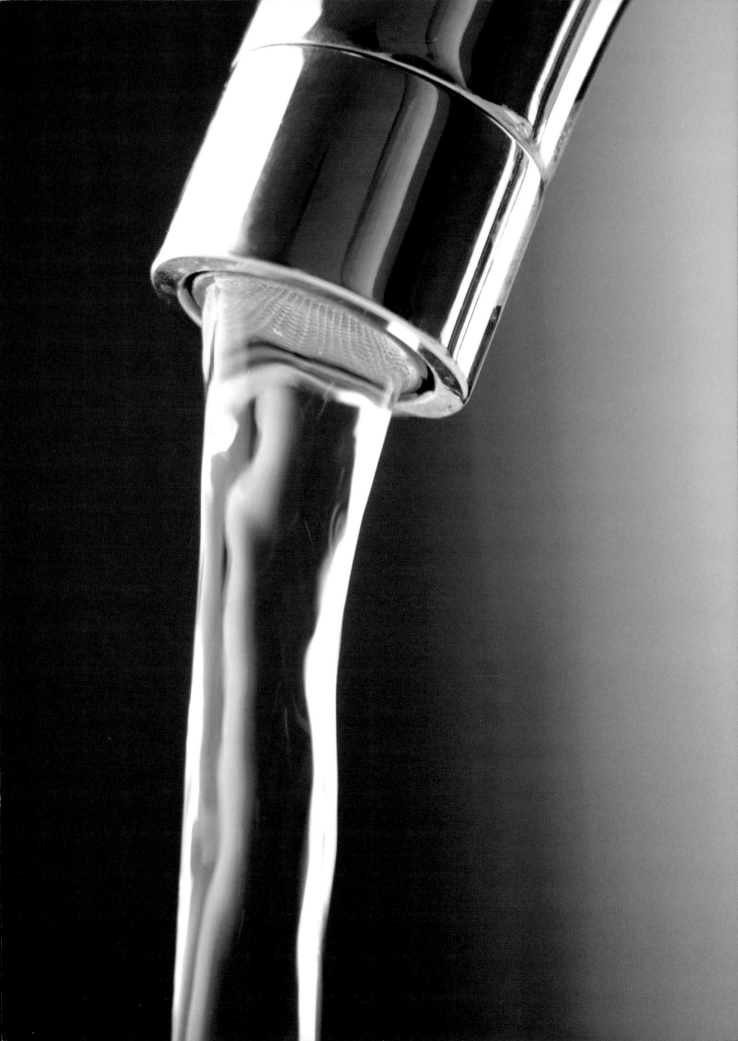

# 3 Simple Projects

**Before you start a major plumbing project, build up your confidence with some simple tasks that improve your existing system.**

# doit FIX A DRIPPING TAP

In an older house you will probably need to turn off the water at the main stopcock (see pages 13 and 48). This is usually under a sink, in a garage or utility room. Check that it really is off by opening the cold tap in the kitchen. When working on an indirect hot water system, you should turn off the valve next to the tank, or even empty the tank.

**1** Remove the tap mechanism from the main body **A**. Make sure you turn the water off at the stopcock or isolation valve first!

**2** With the mechanism out you can see a fibre washer where the mechanism meets the body **B**. If you have a leak from this area, replace that washer. This one is OK.

## Before you begin

Before working on the water supply system you should turn off the supply to the area that you are going to work on. In a modern or recently refurbished house there will be isolation valves on each feed to a tap, WC or any other fitting, so turn off the appropriate one.

## Traditional Taps

These are rare indoors, but are still common outside as they are simple, cheap and robust. They can drip from three locations: around the spindle, from where the tap mechanism screws into the body, or from the spout. If your tap is leaking from around the spindle, see *Fix a Dripping Stopcock*, (page 72). To cure other drips you need to do the following:

A

B

Garden taps should be fitted with a check valve to prevent dirty water and garden chemicals syphoning back into the system if the water pressure falls. You can fit one somewhere in the feed to the tap, but the simplest solution is to buy a garden tap with a built-in check valve. Just unscrew the old one and screw in the new.

**3** Turn it over and remove the old washer. You will probably have to prise it off with a screwdriver **C**. On very old taps there is a little screw holding the washer.

**4** Press a new washer into place and refit the mechanism into the body **D**.

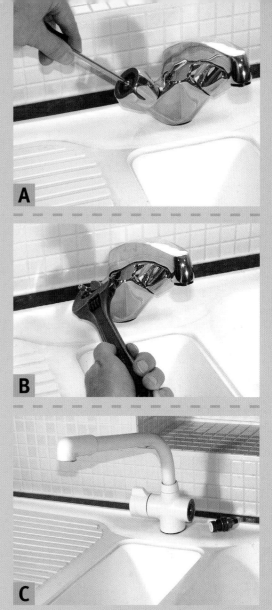

## Modern Taps

Modern taps can drip too. Tap designs vary hugely but most have some sort of push-fit cover you can prise off, and under that a screw to hold the handle **A**.

**B** This tap uses conventional tap washers, so once the handle is off you can remove the mechanism in exactly the same way as the traditional tap opposite.

**C** Even ceramic disc taps come apart in the same way, but having got this far you cannot strip the ceramic cartridge any further. Take it along to your plumbers' merchant and see if they can provide a replacement cartridge. Many designs do use interchangeable cartridges but if you are unlucky you may have to replace the entire tap.

# FIX A DRIPPING STOPCOCK

Stopcocks are just a form of tap and if there is a tiny leak past the tap washer you may never notice. If it gets to the point where you cannot turn off the flow sufficiently to carry out whatever work you have in mind, you will need to replace the washer, attend to the seat, or even replace the whole thing just as with a tap, but with one important difference – you need to turn off the company stopcock outside before you dismantle the internal one!

Stopcocks are rarely used so they rarely wear out and, as they are usually hidden, there isn't the incentive to change them for appearances' sake. It's common to find an old-fashioned model like the one shown, even in a house that has been extensively modernised. This type has a gland to make a seal around the spindle. When the cock was installed, the gland was stuffed with waxed string and the nut tightened to compress the string around the spindle. It will remain watertight for years but once you disturb it by closing and opening the stopcock it may start to leak.

**A**

1 Here you can see the problem, the drip forming under the stopcock **A**, and the potential solution. If the gland packing is sound, then tightening the gland nut may stop the drip **B**. Don't over-tighten though – you still need to be able to turn the handle!

2 To replace the packing, turn the stopcock off and slacken the gland nut. Remove the old packing using a small bent screwdriver or even a bent nail. Wind some PTFE tape round the spindle and push it into the gland, then replace the nut **C**.

**B**

**C**

# doit **REPAIR OR REPLACE A STOPCOCK**

If you need to work on the stopcock itself, you need to go one step further back along the supply system to the water company's stopcock, which is located somewhere near the bounday of your property, usually either by your gate or on the pavement outside. There will be a small access hole covering the water company's stopcock. You need to switch that off before working on your own internal stopcock. If you cannot reach the handle, you need either a stopcock key (right), or a piece of 2 x 1 timber with a notch cut in the end.

Plastic stopcocks with push-fit fittings are simple to fit, but you will need to cut off the old fittings and extend the pipework. In many cases it is easier to replace it with a traditional type that connects to the original pipes.

**1** Make sure that you have turned off the company stopcock, then remove the old stopcock **A**.

**2** Wind some PTFE tape around the olives on the existing pipework and fit the new stopcock. If the old olives won't seal in the new stopcock you will have top cut them off and extend the pipework, but this is rarely necessary **B**.

## STOPCOCK KEY

It is easier to turn off the company stopcock using this key as it provides more leverage and reaches deeper.

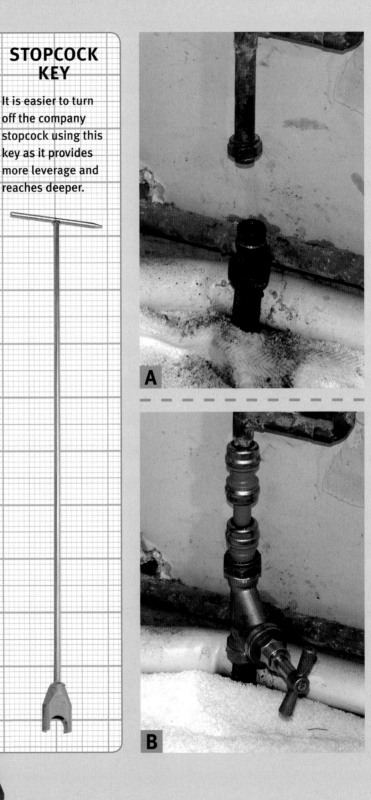

**A**

**B**

**DISMANTLED STOPCOCK**
Stopcocks use the same 'tapwasher' system as old style taps and can be repaired in the same way (see page 70).

# doit FIT A GARDEN TAP

If you are looking for an easy project to begin DIY plumbing, they don't come much easier than this. All DIY stores sell outdoor tap kits which contain a saddle valve to attach to a convenient cold water pipe, some plastic piping, a tap back plate, usually with a built-in check-valve, and a suitable tap.

**1** First, find a suitable cold pipe. Most houses have kitchens at the back and most people want a tap in the back garden so a pipe under the kitchen sink would be ideal. TURN OFF THE WATER SUPPLY.

**2** Screw the valve into the saddle **A**. There is usually a locknut on the valve. If so, screw that up as far as possible so the valve goes right in as shown right.

**3** Fit the saddle backplate behind the pipe and secure it to the wall **B**. It needs to be firm since you are going to put quite a strain on it. If the pipe is clipped too close to the wall you may need to pack it out. If it is on plastic clips you may need packing behind the back plate.

**4** Put the saddle valve assembly on the back plate and tighten the screws. Once the screws are fully home and the pipe has been pierced, you can rotate the valve into position and tighten the locknut to secure it **C**.

## For more information

For more information about other types of taps see *Fit a Bath*, page 81–3, and *Fit a Tap*, pages 106–7.

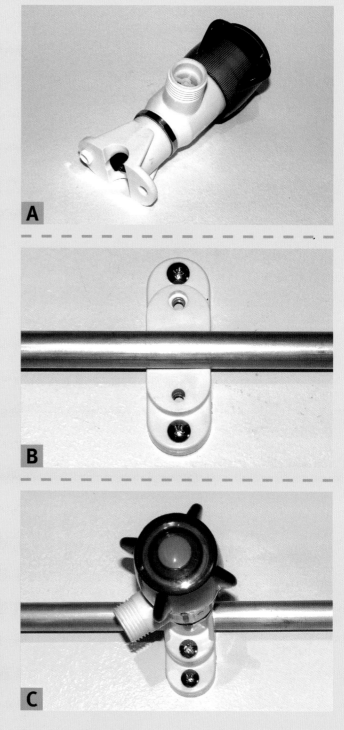

A

B

C

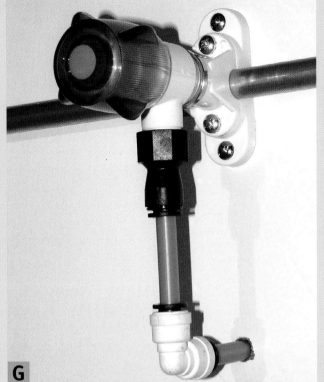

**G**

**D**

**E**

**F**

**5** Drill a Ø12 hole through the wall and counter drill the outside Ø22 deep enough to take the check valve (kits may vary so check yours). Pass a piece of Ø10 flexible hose through it **D**.

**6** Attach the wall plate to the hose and then screw to the wall – you will need to rawl plug it if it is a solid wall **E**.

**7** There will be a fibre washer or O-ring in the kit. Fit to the tap, then screw the tap into the wall plate **F**. If it does not face the right way when it is tight, undo the screws and rotate the wall plate. An alternative method is to screw the tap and wall plate together, then mark out and drill the wall.

**8** Back inside, trim the hose to length and fit it to the valve. Some kits provide elbows like this one, others just rely on the flexibility of the hose. If you do have to bend the pipe, cut it long and make large curves – a tight bend will kink and restrict the flow. Turn the water back on, and go and water the garden! **G**

# doit FIT AN ISOLATION VALVE

If a WC system needs a new ball valve and syphon, it is a good idea to fit an isolation valve first. This means that in future it will be possible to work on it without shutting off all the water in the house.

This job is a good introduction to using compression joints. They come in various shapes and sizes but all compression joints are fitted in the same way. Some plumbers like to smear on jointing compound around the olive, but it isn't really necessary.

Push-fit copper or plastic joints are marginally easier to fit, but bulkier and more expensive.

**ISOLATED WC**
With an isolation valve fitted, future work on the WC is much easier (below).

**1** Hold the valve next to the pipe where you intend to fit it and mark the pipe with a pen **A**. The pipe will slide into the fitting up to the end of the screw thread. Check the water is off before going any further.

**2** When a pipe is fixed to the wall you cannot usually use a pipe cutter, so use a junior hacksaw and have a cloth ready – there is always a little water above the cut that will have to come out. Cut on the mark furthest from the appliance **B**.

**3** Once the pipe is cut, undo the connection to the appliance and remove the cut-off section. With the pipe removed you can file the cut end smooth **C**.

**4** Cut the marked section of pipe with a pipe cutter **D**. These cut cleanly so there is no need to file it.

**5** Remove the nuts from the valve to release the olives. Slide one nut and olive to the other end of the pipe **E**.

**6** Place the valve on the pipe, taking care to check the direction of flow indicated by the arrow on the valve body. Tighten the nut just finger tight at this point **F**.

**7** Slide the other nut and olive on to the remaining piece of pipe, fit it to the valve and re-fasten it to the appliance. Hold the body of the valve with a spanner while you tighten the nuts **G**. Just one turn on the nuts is usually sufficient.

A

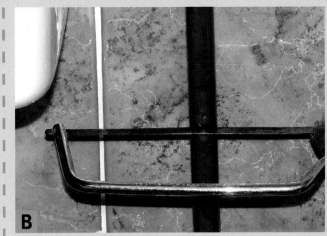

B

C

D

E

F

G

## toptip*

If a compression joint leaks when you turn the water back on, don't panic! An extra eighth of a turn on the nuts is usually enough to seal it. If it doesn't, then turn the water off and loosen the nuts. Wind some PTFE tape round the olives and re-tighten the nuts.

# 4 Bathrooms, Showers, Kitchens and Heating

**Bigger plumbing projects are just more of the same – if you can join pipes and fit a tap, you can plumb a whole kitchen or bathroom.**

# Baths

It is hard to beat a warm, relaxing bath at the end of a tough day. For many families, they are an essential part of the children's bedtime routine.

Baths can be made from a surprisingly wide range of materials including wood (usually teak because it is naturally water-resistant) or stone such as marble, slate or granite. Most baths today, however, are made of metal (steel or cast iron) or plastic (acrylic or GRP).

## 4  Types of bath

**CAST IRON** is the traditional material from which baths were made. In its natural state, cast iron would quickly go rusty, so it is always enamelled. Enamels can be almost any colour, but white is always popular. Cast iron is very strong but very heavy, which can be a real problem if you are fitting it in bathroom at the top of a three-storey house!

**STEEL** is thinner, lighter and cheaper than cast iron. Like cast iron baths, steel baths are enamelled but they are more flexible, which makes the enamel more prone to damage.

**ACRYLIC** baths are moulded from a sheet of plastic. They are lighter than iron or steel and can be made almost any shape or colour. Acrylic is smooth, shiny and stain resistant, but it scratches easily.

**GRP** stands for glass reinforced plastic, which is commonly known as fibreglass. GRP is tougher than acrylic, but can still be scratched by abrasive cleaning products.

**MODERN MINIMALIST**
A typical sleek simple design for a modern house (*above*). It's easier to clean too!

**RETRO-STYLE**
Traditional cast-iron baths (*right*) enhance a period bathroom.

Unless you are building a new bathroom, the first task is usually removing the old bath.

**1** Start by removing the panels. If there are no obvious fittings they are probably held by clips and should just pop off. Beware of new floor coverings though – if someone has tiled up to the panels, they may be hard to remove.

**2** If the house was built in the last 20 years or so there will probably be isolation valves on the hot and cold supplies to the taps. If so, turn them off and open the taps to make sure the water really is off.
   If there are no valves, turn off the supply wherever necessary and check by opening the taps.

**3** Disconnect the pipes leading to the taps, ideally from the isolation valves but you have to use your own judgement if the layout is different.

**4** Disconnect the waste pipe by unscrewing the trap from the bath.

**5** Check its feet. Cast-iron baths are so heavy that they are rarely secured. Other baths should be screwed to the floor and will need unscrewing. Baths are usually wider than the door, so you will probably need to remove its feet and turn it on its side to get it out.

**6** Fitting the new bath should be a simple reversal of removing the old one, though it's rare that the pipes line up exactly. The simplest way to address this is to cut them off and use flexible tap connectors (see page 39). If the existing system doesn't have isolation valves the cheapest, easiest solution is to use a flexible tap fitting with an integral valve **A**.

(*continued on next page*)

A

**B**

**C**

**D**

**7** It is far easier to fit feet **B**, taps, waste outlets and overflows **C** to a bath on its side, rather than after fitting. It is also a good idea to connect the flexible pipes to the taps beforehand; when the bath is in place you can simply reach under and fasten the pipes to the exiting supply.

Check out fitting instructions for the side and end panels as well. If you have to screw clips in place, this is also easier with the bath upside-down. Place it on something soft to avoid scratching.

**8** This steel bath is in a modern house with chipboard floors, and chipboard is not ideal for supporting heavy loads on small areas. Instead of using the adjustable feet, the legs rest on wooden boards that run across the joists supporting the floor. Note the small piece of plywood used as a 'shim' under the nearest leg to level the bath **D**. You will probably also need a flexible waste pipe **E**.

## toptip*

Do not try lifting a cast-iron bath single-handed. If you put your back out at this point, you won't even have a hot bath to soak it in!

**E**

A

B

## Problems and Solutions

DIY is rarely straightforward, especially in old houses where strange things have been done by previous owners. This interesting example demonstrates some problems and solutions.

This bathroom is in a poorly-constructed rear extension to an old terraced house. Bath taps are still made to the UK ¾in standard and bath tap fittings are designed to fit Ø22 supply pipes, but in this case Ø15 pipes were used. To get round this the original installer had used an adaptor to connect the ½in taps to Ø15 pipe.

**1** One way to solve this problem is to keep the original adapter, fitting new fibre washers if necessary, and connect a short length of Ø15 pipe as shown. To join to the supply, use a connector with Ø15 compression joints on both ends instead of a conventional flexible tap connector **A**. An alternative, neater solution is a proper ¾in to Ø22 flexible tap connector as shown above, left, but you will also need to fit a Ø15 to Ø22 adaptor to the supply pipe **B**.

**2** When metric waste pipes were first introduced they were simply the old inch sizes translated into millimetres. 1½in (inside) became Ø43 (outside) and this was used here, buried in the floor screed and not easy to change. Modern waste pipes are a more rational Ø40. Uni-compression waste fittings are designed to open up to 43mm and squeeze down to 40mm when tightened, but push-fit fittings and flexible waste pipes must be the right size. Left, a uni-compression 135° elbow has been used to connect the differing sizes **C**.

C

## toptip*

Even if you plan to move the waste pipe and fit a new trap, leave the old one on until you are ready to replace it, to keep foul smells to a minimum.

# Washbasins

## Types of Washbasin

Like baths, washbasins are made from a variety of materials. Ceramic is the most traditional material, and is strong and easy to clean. Ceramic basins are heavy, so they are usually mounted on a pedestal. As well as supporting the basin, the pedestal hides most of the plumbing. Small basins in en suite shower rooms and downstairs WCs are more likely to be wall hung. If the walls are plasterboard, it is easier to fix a corner basin securely than attach one to a single wall. With no pedestal to hide the trap and pipes, you should consider the design of the trap and type of pipework you use.

Acrylic basins are lighter and cheaper but not as rigid. They are normally mounted in a vanity unit that provides all-round support, hides plumbing and gives storage space.

GRP can also be used for basins, but as the extra rigidity it offers over acrylic is not as important as for baths it is less common.

Corian is an acrylic resin that can be moulded in a variety of objects. It is not common in domestic use, but is sometimes used in vanity units. It is often used to make worktops, and is used in hotels to construct seamless rows of basins in a washroom. It is not common in domestic use, but it is sometimes used in vanity units.

Various stone basins are produced for the deluxe end of the market, along with bronze, stainless steel and wood, but you won't see them in normal DIY stores and plumbers' merchants.

**DUAL PURPOSE**
A vanity unit like this encloses all the unsightly plumbing and provides shelf space and storage.

# doit FIT A WASHBASIN

The precise method of fitting a washbasin varies according to type. If you are fitting a new one, the manufacturer's instructions are a good starting point, but here are some general tips that will help.

**1** If you are fitting a wall-hung or a pedestal-style basin, you need an assistant to hold the basin in place on top of the pedestal whilst you mark through the mounting holes for the basin. Use a spirit level to make sure the basin is level before marking. Drill and plug the wall.

**2** Fit the taps, waste outlet and overflow before you fit a pedestal basin in place. They can be hard to reach afterwards. The photo shows how hard it is to reach the nut that secures the tap once the basin is fitted **A**.

Some taps, such as the example shown, have copper tails fitted. Others are threaded so you will need to fit copper or flexible tails. This is easier to do before the tap is in place.

**3** If you are using rigid pipework, you will need help to hold the basin in place while you make up the pipework **B**. You could place the wall screws loosely at this point, but do not expect them to support the weight of a ceramic basin without a pedestal. When the pipework is complete, ease the pedestal into place and tighten the screws.

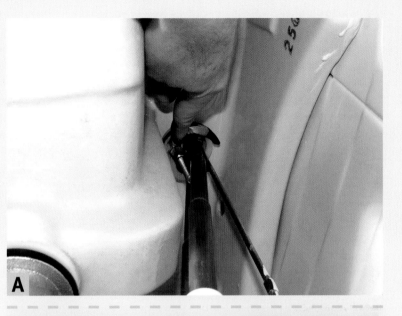

## The Vanity Unit Problem

There is not usually much difference between fitting a vanity unit and a washbasin, but it may be more interesting. This unit is actually an old washstand with the top cut out to take an acrylic basin. It looks right in an old cottage but presents some problems.

**1** The most obvious problem is that over-enthusiastic use of aggressive cleaning materials has damaged the soft gold plating of the waste outlet so it looks permanently dirty **A**.

**2** More serious, though, is that due to limited room inside the washstand, a shallow P trap was used rather than a more conventional S or bottle trap **B**. This would not normally matter much, but the basin is a long way from the stack and the waste pipe connects to the bath waste. The syphoning action resulting from emptying the bath draws water out of the basin trap and allows foul smells into the bathroom.

**3** Removing the old trap should be easy but here it was glued up with silicon sealant **C**. To prevent the outlet from rotating, drop a pair of screwdrivers down the plughole and use something as a lever between them **D**.

The obvious solution is a new outlet and an anti-syphon trap (see page 65), but the new trap is too long vertically and too short horizontally to meet the existing waste pipe. The answer is to cut the screwed section of the waste outlet so the trap fits higher than normal. If you do this, take care to cut straight, and smooth the cut end afterwards. The trap seals to the outlet by pressing an O-ring on to the end of the pipe **E**. If it is rough or crooked it may leak.

**4** The original waste pipe was cut back about 100mm to make room for a straight coupling and a short section of new pipe **F**.

**5** It took an hour or so but the result is a cleaner looking, fresher smelling bathroom. Well worth the effort **G**.

A

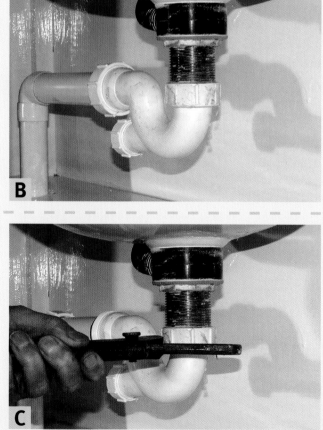

B

C

## BIDETS

These are almost always made from ceramic because it is strong and hygienic. In terms of fitting, just think of it as a washbasin on a very short pedestal!

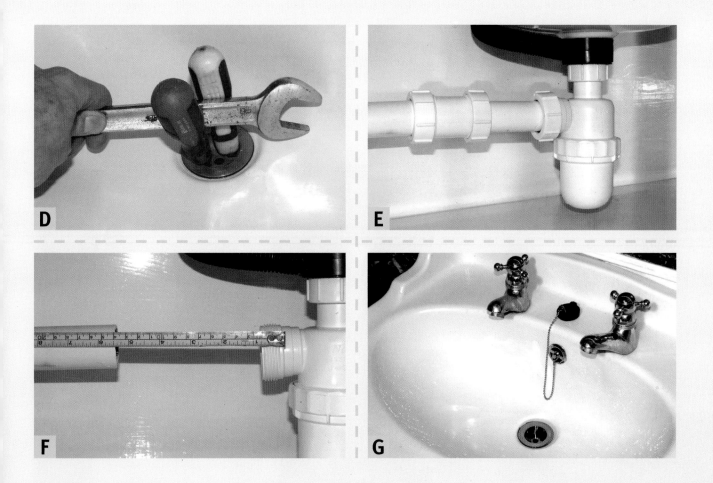

D

E

F

G

# WCs

## Types of WC and Cistern

Apart from stainless steel 'anti-vandal' WCs in public buildings and the occasional 'designer' piece, WC pans are almost always ceramic, chosen because ceramic is sturdy and hygienic. Cisterns are more varied.

Cisterns store about 7–9 litres of water which can be dropped suddenly into the WC to flush it. They were developed when water supplies were erratic and pressures were often too low to provide a good flush. They are still in use in homes in the UK, though mains pressure flushing it is becoming more common in commercial premises.

The first WC cisterns were mounted high on the wall, close to the ceiling, to increase the force of the water falling into the pan. Modern designs have more efficient valves that can create sufficient flushing action without such a drop.

Low-level cisterns set about a metre above floor level have been standard for many years, but they are now being superseded by close-coupled cisterns that actually rest on the back of the WC. Old, high-level cisterns were made of cast iron, but you are unlikely to see a cast-iron cistern unless you are restoring an old house. Later cisterns are generally ceramic, though some cheaper, low-level cisterns are plastic. Plastic cisterns are generally wider and slimmer than ceramic. This is used as a marketing ploy: though allows the WC to be fitted slightly closer to the wall, the design evolved as much to ease the strain on the plastic and wall fittings by reducing the overhang.

Close-coupled cisterns are always ceramic and are usually bought as a set along with the WC. Close-coupled cisterns look tidy and are slightly cheaper to fit as there is no flush pipe to attach. The WC must be positioned accurately to ensure that the cistern fits against the wall when placed on the pan. The chief advantage of close-coupled cisterns is that the weight of the cistern (and water) is supported by the floor rather than hanging on the wall, which in a modern house is often too insubstantial to support it.

The water level in a cistern is controlled by a float-operated valve. These are generally spherical and fitted to the end of an arm which operates the valve, giving rise to the old name 'ball cock'. Newer valves are operated by a cylindrical float, which rises and falls vertically. This is particularly

**DOWN LOW**
Low-level cisterns are fitted about a metre from the floor (below), while close-coupled cisterns rest on the back of the WC unit (*below left*).

common where the entry point for the water is in the base of the cistern rather than the top. They don't work any better, but look neater because they avoid having a pipe up the side of the cistern.

## WC Drains

Almost all WCs connect to a soil stack using Ø110 pipe. This should be fitted with a slight downward slope, and to make this easier branch fittings are made to a 92.5° angle. If you are fitting a branch to a stack, use an access branch. These have a removable cover opposite the entry point for the branch, so it is easy to use a drain rod if it gets blocked.

## Macerators (Saniflow)

Sometimes though Ø110 pipe is not really convenient, especially when a WC is being fitted a long way from the stack. This is often the case when en suite bathrooms are added, or when a house is converted to flats; the stack is traditionally at the back of the house, which is not much use if you put a WC in the front bedroom. The answer to the problem is a macerator.

A macerator is designed to fit behind a WC connected to the waste outlet. Inside is a rotor that resembles a food processor and reduces solid waste to a slurry, plus a pump that pushes it along a conventional Ø40 waste pipe. The pump ensures that the system will work even when there is a long horizontal section of waste pipe, for example, from the front bedroom to a soil stack at the back.

Macerators are also useful if you are fitting a bathroom in a cellar. In this situation you may well find that the public sewer running under the street is actually higher than your bathroom, so drainage by gravity isn't going to work. Macerators able to lift waste to a height of 5 metres are common. The larger ones can accept input from basins, baths and showers as well as the WC, and will pump the lot up to the soil stack or drain.

When fitting waste pipes for a macerator, aim to have as few bends as possible and avoid sharp bends completely. Use a pair of 135° elbows rather than a single 90°.

There are many types and sizes of macerator from several makers, Saniflow is one of the better known. Like other trade names, it is often applied to other makes. The model shown here is ideal for small cloakrooms. The macerator is built into the WC base and the flush system works from mains water pressure, so no cistern is needed. Whatever type you choose, follow the maker's installation guide carefully.

## CAUTION
## SANIFLOW
All macerators (not just Saniflow) need an electrical supply, and bathroom wiring is subject to strict regulation. It is possible to do it yourself, but contact your local authority's building control department before starting work. For more advice on electrical work, consult an electrician.

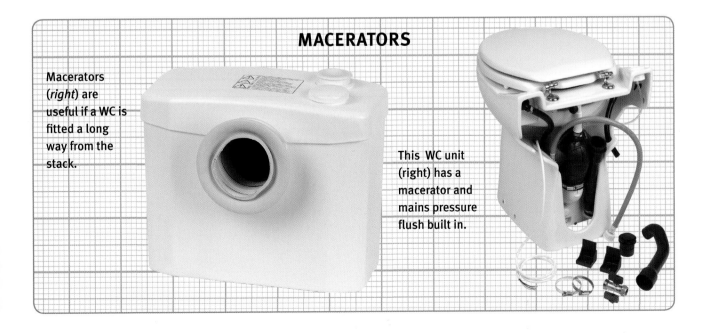

## MACERATORS

Macerators (*right*) are useful if a WC is fitted a long way from the stack.

This WC unit (right) has a macerator and mains pressure flush built in.

# doit FIT A WC AND CISTERN

WC pans are screwed to the floor, and a solid floor will need Rawlplugs. Cisterns are screwed to the wall, and the only potential problem is getting them in the right place. Close-coupled cisterns are the easiest to deal with.

## Close-coupled Cisterns

**1** Assemble the flush mechanism and fit the rubber gasket to the outlet.

**2** Position the cistern on top of the WC pan and slide the whole thing into place, so that the cistern is against the wall. If the waste pipe is already there, make sure you line up with it. The pipe may be lengthened or shortened if necessary, but bending it if you are out of line is tricky.

**3** Mark the screw holes, then remove the WC and cistern while you drill and plug.

## Low-level Cisterns

Low-level cistern systems are slightly more flexible because the flush pipe is usually longer than necessary, so the height of the cistern and distance of the WC from the wall are not so crucial. This is particularly useful if you are fitting a new WC to an existing waste pipe. It's much easier to cut a flush pipe to suit than it is to move a waste pipe. If you are starting from scratch, follow the manufacturer's guidelines regarding heights and distances.

Supply connections to cisterns are just straightforward Ø15 pipework. Don't forget to fit an isolation valve nearby to make any future work on the cistern simple.

**1** Connecting the flush pipe on a low level system is very easy. The cistern end uses a form of compression joint. You just slide it in and tighten the nut. The WC pan end uses a 'rubber' (actually a form of plastic) sleeve as shown, or a push-in 'rubber' bung **A**.

**2** Waste pipe connections are equally simple providing the pipes line up. Designs vary, but in general the WC waste outlet fits into a socket on the end of the pipe and is sealed by some form of gasket **B**. Flexible connectors are available for awkward situations.

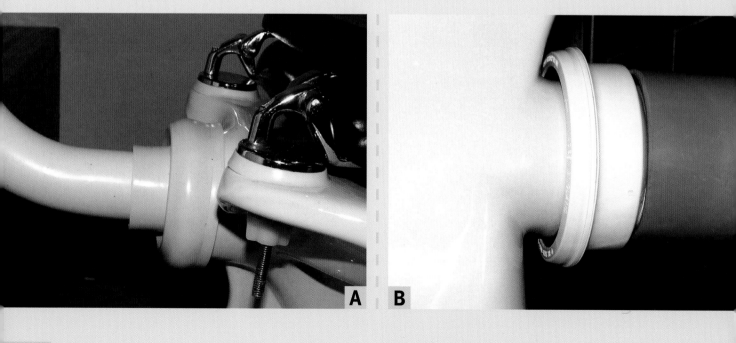

A   B

# Troubleshooting Cisterns

**SMALL-SPACE SOLUTIONS**
Modern systems mean that it is possible to fit a WC in almost any space. (*above*).

There are two separate valve systems in a WC cistern, one to let the water in and one to let it out again. Either can fail.

## Cistern Overflow

Float operated valves are very reliable, but eventually they do fail. In hard water areas the valve may get so coked with limescale that it refuses to open or close properly. Sometimes dismantling it and soaking the parts in limescale remover is sufficient to fix it, but given the low cost of a new valve it is hardly worth the effort.

Early ball floats were made of thin copper which eventually corroded, sprang a leak and sank. Modern ones are made of plastic, so sinking is rare but can happen where a badly adjusted valve has the ball rubbing on the side of the cistern until it wears through. If the rest of the valve mechanism is fine you can simply unscrew the ball and buy a new one.

If a ball sinks or the valve jams open, the cistern will overflow. Until recently, cisterns were always fitted with an overflow pipe, on the opposite side to the water inlet. To avoid having to make left and right hand models, the inlet and overflow holes are identical.

A 90° overflow connection (see below) is used when the overflow is connected to the top of the cistern. If it is a bottom filling cistern then the overflow also comes straight out of the bottom. Ideally, such overflows should be clearly visible outside the house so you quickly become aware of the problem. If it is tucked away out of sight you could waste a lot of water before you notice it.

More recent syphons and flap valves have an internal overflow so that water trickles into the WC pan. This is obviously easier to install, but you may not notice a tiny trickle. If the surface of the water in the WC looks even slightly rippled long after it has been flushed, investigate the level of water in the cistern.

The picture of a low-level cistern (see page 88, left) illustrates a less-common situation. This WC is fitted at the back of a large building on a sloping site, so if you were to drill through the back wall for an overflow it would emerge below ground level! The plumber has therefore connected the overflow to the flush pipe, achieving the same result as an internal overflow.

## OVERFLOW FITTING

Cistern overflow fittings are right-angled to point backwards to, and lead out of, the wall.

## Syphons and Flap Valves

There are two different methods of letting the water out of a cistern: syphons and flap valves. Syphons have been used since the nineteenth century and, though the design has changed a lot, the principle is the same: operating the handle lifts some water into the flush pipe and lets it fall. Once the flow has started, more water is syphoned up until the level in the cistern falls below the bottom of the syphon, then it stops.

Very old systems had a bell-shaped syphon that was bulky and noisy but more or less indestructible. Modern ones have an inverted U-shaped tube (see also page 62),

## doit REPLACE A BALL-VALVE

1 Turn off the the isolation valve on the supply to the cistern **A**. If there isn't one, see *Fit an Isolation Valve* on page 76. Flush the WC to empty the cistern and make sure the water is really off (the cistern doesn't refill).

2 Undo the union (joint) connecting the pipe to the valve **B**. Some water will drip out at this point so have a cloth handy.

3 Undo the nut holding the valve into the cistern. Take it right off, and then remove the valve from inside the cistern.

4 The ball and the valve are usually sold separately, so screw them together before fitting the valve **C**. Try the assembled valve in place – you will probably find the ball touches the back of the cistern.

5 Some valves have bent or adjustable arms. If not, you will have to bend it yourself to suit your cistern. A vice is ideal but a mole wrench will do **D**. Remove the ball first and be careful not to damage the screw thread. Bend it a little at a time and keep checking.

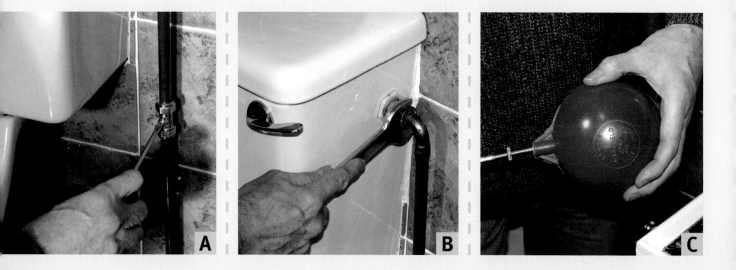

A     B     C

with one end open near the bottom of the cistern and the other connected to the flush pipe. The open side has a piston inside which is lifted by the handle to start the flow, and the piston has a flap valve built in (or arranged as a bypass) so that once the first lot of water goes over the U-shaped tube and starts to fall, more water can pass straight through (or past) the piston. In hard water areas the flap valve in the syphon can get clogged with lime. Dismantling it and soaking in limescale remover might restore it.

More common are problems with the connection between the handle and piston. Since syphons and cisterns come in various shapes and sizes the mechanism has to be adjustable. There is usually an arrangement of levers and links that can be quite flimsy. If the WC won't flush, take the lid off the cistern and examine the mechanism inside. It's usually obvious what the problem is. Adjusting the lever may help, or if you have a lever with several holes, try connecting the link to a different one. If it's a plastic lever it may have broken. Fortunately, they are readily available, cheap and easy to replace.

Eventually, there comes a point when tinkering is no longer enough. Replacing a ball valve is easy. A flap valve or syphon is more work, but not too complicated.

**FLAP VALVES**
These are held in place by a simple twist and click bayonet fitting (*left*).

---

**6** When the ball valve will operate without touching the cistern or the syphon you can fix it in place. Hold the valve with a spanner or pipe wrench whilst you tighten the nuts **E**. Before you re-fit the union, check if it has a fibre washer or O-ring that needs replacing.

**7** Turn the isolation valve back on and watch the level in the cistern carefully. It will normally stop before the correct level, but be ready to switch it off if it doesn't. If the level is nearly right, try the flush **F**. If it works OK leave it, but if the level is much too low the WC will not flush properly. Some valves are adjustable. If it isn't you'll have to bend the arm up or down a little.

**8** Turn off the isolation valve, remove the split pin holding the arm in the valve, then you can remove the arm without having to take off the whole valve.

**toptip***

**Fibre washers on unions go hard, and O-rings deteriorate after a few years and don't seal well. Before dismantling any valves, have a packet of assorted O-rings handy.**

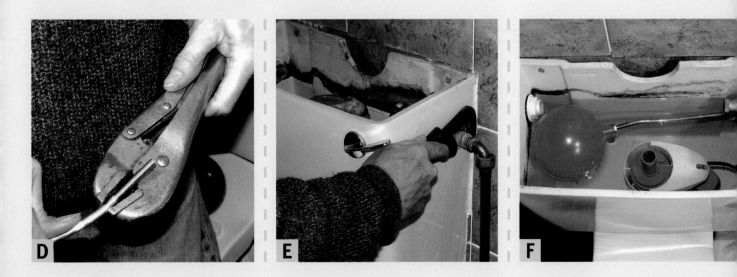

D

E

F

# REPAIR OR REPLACE A FLAP-VALVE

Modern flap valves allow precise control of the amount of water used in a flush, and most are now 'dual-flush' which can reduce your water bill. Unfortunately they do not seem to be designed with durability in mind.

In hard water areas a build-up of limescale on the rubber disk in the valve can prevent it sealing properly. This will result in a continuous dribble of water into the WC pan. The first step is to check the float valve (ball cock). It may be that the level in the cistern is too high and that water is running down an internal overflow, but if that is not the case you should investigate the valve itself.

A

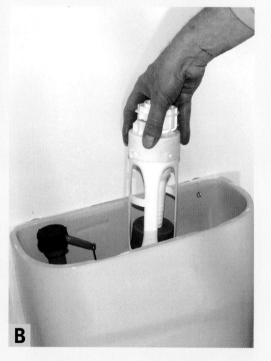

B

**REMOVING A FLAP-VALVE**
Flap valves use a bayonet fitting for easy removal and replacement.

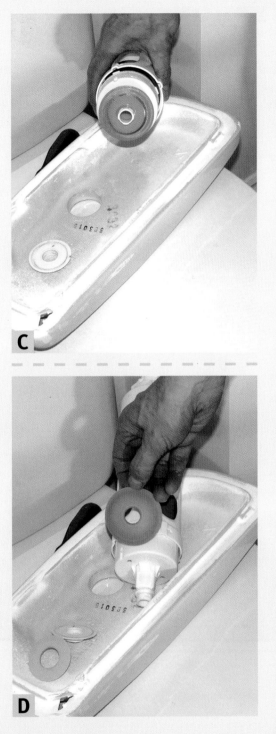

**1** Turn off the the isolation valve on the supply to the cistern **A**, flush the WC to empty the cistern and make sure the water is really off (the cistern doesn't refill).

**2** If you have a ball valve, remove the arm (see *Replace a Ball Valve*, page 92) as it will get in the way. Torbec-style valves cause fewer problems.

**3** Flap valves are fitted into the base of the cistern using a 'bayonet' system. Grasp the valve firmly with both hands, avoiding any of the most obviously flimsy parts and twist it anti-clockwise. Right at the point where you think it is about to fall to pieces it will (usually) come loose **B**.

**4** Turn the valve upside down and inspect the rubber disk **C**. If it is damaged you are going to need a new valve because no-one seems to sell replacements. If it is coated in limescale, carefully prise the disk off with a small screwdriver and remove the scale with some gentle sandpapering **D**.

**5** Re-assemble the valve, replace the ball cock if necessary and turn the water back on. With luck you've fixed it, but if it still leaks, or the problem lies in some other part of the valve, you are going to have to replace it.

**6** The bayonet fitting on valves is common but not universal, so remove the defective one and take it with you when you go to buy a replacement. If you can find a new one with the same fitting, simply pop it in. If you cannot, you will have to remove the cistern and replace the lower end of the bayonet fitting too. This isn't particularly difficult, but it will involve some heavy lifting if you have a close-coupled ceramic cistern.

Replacing a syphon, or fitting a flap-valve system in place of a syphon, are similar processes. These are combined on the following pages.

**1** Turn off the water supply, empty the cistern and remove the float valve (see *Replace a Ball Valve*, page 92) **A**. Some water will remain in the bottom of the cistern. Scoop it out with a cup or small jug, then mop up the last drops with a cloth. There is a good chance any that remains will end up on your feet when you remove the cistern!

**2** Remove the screw holding the cistern to the wall **B**. A close-coupled cistern like this one also has two wing nuts holding the cistern on to the WC that you must remove from underneath.

Low-level cisterns **C** may also have a bracket underneath to take the weight. The cistern just rests on them so there is nothing to undo, but you will have to remove the flush pipe. This is held by a screw like a large compression joint on the cistern and usually a simple rubber push-in seal to the WC. Old WC flush pipes may be held by tape, cement, putty, bandages and layers of paint!

**3** Lay an old towel or similar on the WC lid, lift the cistern off and place it on its back on the WC.

**4** Close-coupled cisterns have a soft rubber gasket to make a seal between the cistern and the WC pan **D**. If re-used this gets squashed and will rarely make a good seal, so remove it and replace with a new one.

**5** The syphon is held in place by a large nut. To remove it, you wll probably need a Stilson or Footprint wrench.

In the case of close-coupled cisterns this nut also holds the coupling. If it is old and rusty, it should also be replaced.

**6** Check that the length of screwed pipe protruding from the syphon/valve is similar to that you removed **E**. If it is too long saw the spare bit off, or you will have problems connecting to the WC or flush pipe.

**7** Whether you are fitting a syphon or a flap-valve, just reverse the dismantling process. Make sure the syphon is the right way round to connect to the operating lever. With flap-valve systems the bayonet fitting must be aligned so the valve will face the right way in the cistern. Test fit it before you put everything else back together.

Some syphon/flap-valve replacement kits come with a large plastic spanner to tighten the joint **F**. If not, use a wrench, taking care not to over-tighten the plastic screw thread.

This cistern has been completely refurbished with a new ball cock **G** and pushbutton-operated dual flush system **H**.

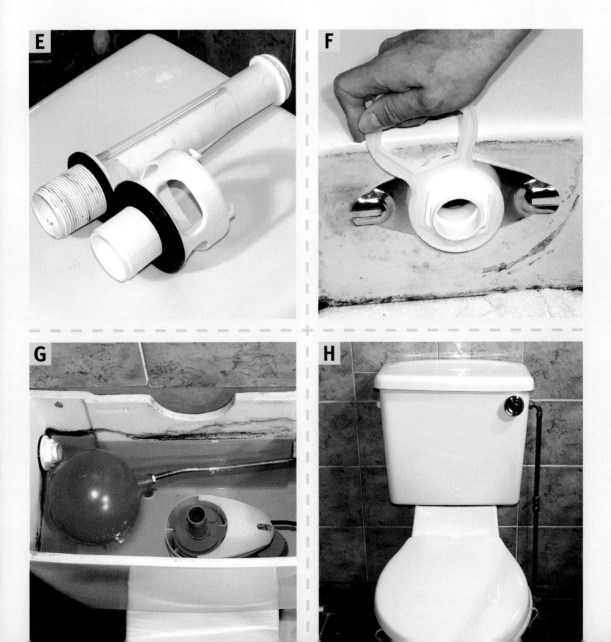

97

A shower is an invigorating way to start the day. It is also quicker, cheaper, more hygenic and uses less water than a bath. Showers fitted over a bath connected to the bath mixer tap don't need any extra explanation. Fitting a separate shower is a little different though.

## Shower Bases

Shower bases (or trays) can be made from a wide range of materials, but cast-acrylic resin is popular for larger sizes and moulded acrylic for smaller ones. There is a fantastic range of sizes and shapes – 900 x 900mm is about the minimum for an adult to shower in comfort – but smaller ones are acceptable for occasional use.

Showers designed to fit in corners can be quadrant shaped, or basically square with a corner cut off – a form of pentagon. If you buy a shaped base like this it's a good idea to buy the enclosure to suit it, but generic surrounds or shower curtains will suit most rectangular or square bases.

Some enclosures provide complete all-round protection from leaks, but many showers are set against a wall, or installed in a corner against two walls. Preventing water running between the shower base and a wall is a major issue.

The most basic design of base has flat or rounded sides. The ideal way of fitting these is to apply a good squirt of silicone sealant between the base and the wall before tiling, then tile down to the base. Finish off with a fillet of sealant between the tiles and base. Alternatively, tile the wall first then fit the base with sealant between it and the tiles.

A much better solution is to buy a base with 'upstands' – vertical edges about 12mm high – which you can tile over. Bases are available with upstands on one, two, three or four sides. Some enclosures do not work with upstands, however, so remember to check with the supplier before buying.

Of course, if your bathroom is big enough you will not need an enclosure at all. One problem common to all shower bases is how to fit them low enough, yet still be able to get a trap and waste pipe underneath. A wet room has the same problem on a larger scale: for this you need a waterproof false floor with drainage underneath.

If you are working on a suspended wooden floor you may be able to fit the trap and waste pipe in the space under the floor.

On a solid floor you normally need to make a plinth using timber and waterproof plywood, though if you are building a ground floor extension for a shower room then you could work out roughly where the trap will be and leave a gap in the floor screed.

A simpler method, especially for small showers, is to use an acrylic base with feet rather like those on a bath. These allow you to level a shower even on an uneven floor and create space for the trap.

## Anatomy of a **Shower Base and Plinth**

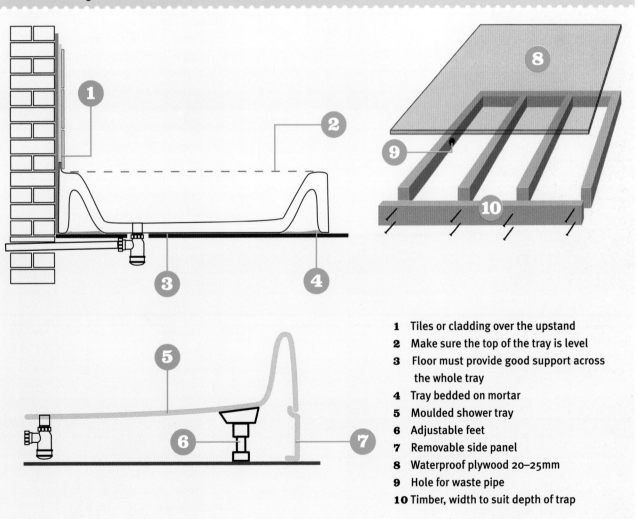

1  Tiles or cladding over the upstand
2  Make sure the top of the tray is level
3  Floor must provide good support across the whole tray
4  Tray bedded on mortar
5  Moulded shower tray
6  Adjustable feet
7  Removable side panel
8  Waterproof plywood 20–25mm
9  Hole for waste pipe
10 Timber, width to suit depth of trap

## Shower Units

The simplest method of supplying water to your shower is just to connect the mixer unit to the normal hot and cold supplies, but this can be a problem in older houses where the water supply to the bathroom comes from a cistern in the loft. In these cases, the showerhead may only be about 500mm below the water level in the loft, which will not provide much pressure. A power shower has an electric pump built in to the control unit to increase the pressure.

An electrically heated shower unit needs only a cold water supply (and electricity), which makes them popular in en suites, flat conversions and so on because the pipework is simpler. They have the added advantage of providing a hot shower even if the central heating / hot water boiler fails.

If installing an electric supply in your bathroom doesn't appeal (see box) a Venturi shower might solve the problem, providing your cold water is at mains pressure. In the Venturi unit the flow of cold water at mains pressure creates a partial vacuum that sucks hot water from the tank even if the hot water pressure is quite low. It won't work if the cold supply is via a tank in the loft, but it might be easier to re-route the cold supply than to get involved in electrical work.

**PRACTICAL LUXURY**
A shower is not only quick and refreshing, but also uses less water and energy than a bath.

# Anatomy of a **Venturi Shower**

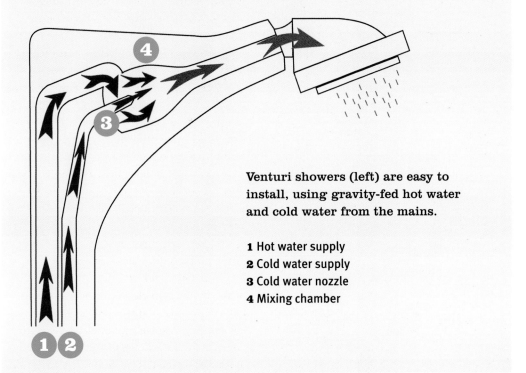

Venturi showers (left) are easy to install, using gravity-fed hot water and cold water from the mains.

**1** Hot water supply
**2** Cold water supply
**3** Cold water nozzle
**4** Mixing chamber

## CAUTION
## ELECTRICITY IN THE BATHROOM

■ Electricity and water are a dangerous combination and new building regulations (England & Wales) insist that any electrical work in bathrooms and kitchens is carried out by an electrician qualified to certify his work, or if done by an amateur is inspected by the council building control department. The building warrant system in Scotland is different but such work still has to be checked.

■ Replacing an existing shower unit is allowed but if any new wiring is needed, contact your local authority's building control department before starting work. If there is any doubt about the safety or suitability of the existing wiring for the new shower seek expert advice.

■ Building Regulations are designed to make our homes safer. Failure to comply could be fatal and at the very least will complicate matters if you decide to sell the property.

## SHOWER WASTE

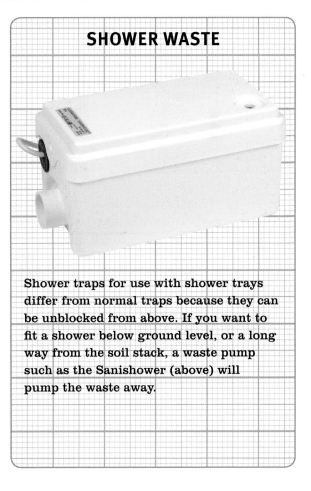

Shower traps for use with shower trays differ from normal traps because they can be unblocked from above. If you want to fit a shower below ground level, or a long way from the soil stack, a waste pump such as the Sanishower (above) will pump the waste away.

# Kitchens

**EVERYTHING BUT THE KITCHEN SINK**
The clean lines of a modern kitchen may conceal
a great deal of plumbing.

Plumbing in kitchens is essentially the same as anywhere else, but it tends to be more complicated because we try to squeeze so much into a small space. The photograph (above, right) is not unusual, but at first sight it's a nightmare.

Starting at the bottom, the blue pipe is the rising main. Fitted to the main is the stopcock and above that is a drain cock. Two elbows and a short piece of pipe lead to the back of the cupboard, where a tee joint sends water to an appliance valve for a dishwasher on the left. Right of this, another tee joint directs water up to the cold tap. Further right, hidden behind the waste pipes, another tee joint pointing down leads to a

garden tap, and the pipe going out of the photograph has yet another tee joint that supplies a washing machine and the boiler.

Above the cold water supply the pipe entering the back of the cupboard is the hot water supply coming from the boiler. That is a bit less complicated with tees leading to the hot tap and washing machine.

In front of everything is the waste system based on a P trap below the sink. The sink is a one and a half bowl design so there is another outlet on the far right leading into the trap. There are also waste hoses from the dishwasher and washing machine, and right at the back a connection to the overflow from the sink.

# Types of sink

As with bathroom fittings, kitchen sinks can be made from many different materials:

**STAINLESS STEEL A** is one of the most common. It's called 'stainless' because, unlike ordinary steel, it doesn't rust. It does stain, however, especially if you empty tea and coffee into the sink, but the stains clean off easily. if you leave plain steel or cast iron items such as a wok or frying pan in contact with stainless steel for a few hours you may get superficial rust, but that too cleans off quite easily. Professional kitchens use stainless steel, which is a pretty good recommendation for its use.

**OTHER METALS** such as copper are expensive but might suit some period kitchens. Copper repels most bacteria but it is harder to keep clean than stainless steel.

**CERAMICS B** have been used for sinks for generations, and modern designs are still being made. Ceramics are heat resistant and easy to clean, though they can chip if you drop heavy things into them.

**ACRYLIC SINKS C** are bright and shiny, and available in a wide range of colours. They are not good for heavy use though. Hot pans and oven dishes will melt them, and cutlery will scratch them.

**CAST RESIN** sinks are more durable than acrylic, but difficult to keep clean. They have a matt finish but tend to stain easily, especially the white ones.

**CHEAP PLASTIC SINKS** are commonly fitted by builders who are anxious to keep costs down. They stain and scratch easily, so you will probably want to replace them within a year or two.

# Fit a Kitchen Sink

The majority of kitchen sinks are designed to fit into a worktop from above and have a small lip that forms a seal on to the worktop surface. This is essential if the worktop is a standard MDF/Chipboard substrate with a melamine laminate surface, because the cut edges are impossible to waterproof. Top mounted sinks are provided with a 'rubber' gasket to make a seal and some form of hold-down devices to clamp them tight to the worktop.

The disadvantage of top-mounted sinks is that you cannot sweep debris into them easily from the worktop. If the worktop is stone, teak, steel or anything else that does not have porous edges you might prefer to fit a sink beneath it. Under-mount sinks are designed differently; it isn't just a case of putting a conventional sink under the worktop, so you need to decide before you buy. Whichever type you opt for invariably comes with a template for cutting the worktop, plus fitting instructions.

It is possible to have cast-resin sinks bonded to a resin worktop so there are no visible joins at all, but you need a specialist kitchen fitter to make the worktop for you. In professional kitchens, stainless steel sinks are welded to stainless steel worktops, and keen cooks sometimes have 'pro' kitchen fitters install a system like this at home.

The deep 'Belfast sink' has had a bit of a revival recently but it is awkward to combine with a modern kitchen. This type of sink was never intended to be fitted under a work surface so if you do, it will be far too low for comfortable use. The traditional method was to position them so that a woman (it was always a woman in those days) who was standing upright at the sink, with her hands clenched, could touch the bottom with her knuckles. As sinks then were used for many different purposes, including doing the laundry and bathing children, this was essential to avoid backache.

Belfast sinks are heavy. They can be supported on heavy duty brackets screwed to a sturdy wall or even on brick piers, but timber frameworks are more common. A flimsy chipboard kitchen unit isn't really strong enough.

**VIEW FROM BENEATH**
Most contemporary designs of kitchen sink are fitted into the worktop from above.

---

**BELFAST SINK**
The Belfast sink (*below*) is too heavy to fit into most modern kitchen units.

---

## 5 Types of Tap

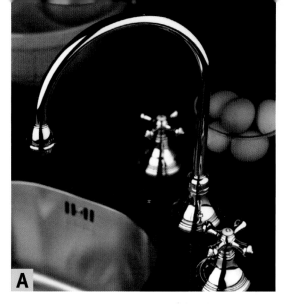

**A**

**SINGLE-SPOUT TAPS** are rare in kitchens these days, and even the wide mixer taps designed to fit a pair of tap holes are becoming less common. Three-hole taps – that's one hole for each tap and one for the spout such as the one shown here **A** – are even more rare.

**MONOBLOC TAPS B** that fit in a single hole in the sink are almost ubiquitous.

**CHROME-PLATED TAPS** complement the stainless steel sinks to which they are usually fitted, but plastic-coated taps in various colours are available to complement plastic sinks. For retro-styled kitchens there are a variety of special effects including antique bronze. There is also a huge range of styles too, but unless you love cleaning (or rarely cook and make a mess in the kitchen), a simple design in smooth chrome finish is hard to beat.

**CHEAPER TAPS C** are still made using the traditional tap washer mechanism, which needs a couple of turns to go from closed to fully open. Quarter turn taps with ceramic disc seals are becoming more popular. Models with long levers that can be elbow-operated are useful if you have dirty hands. They are also more hygienic, which is why they are used in hospitals.

**B**

**FILTER TAPS D** are designed to provide better-tasting water. Though water from any of the UK water companies is guaranteed safe to drink, in some regions it doesn't taste particularly good. Filter taps filter water for drinking and cooking using cartridges similar to those used in filter jugs. The model shown here is a single-hole design. Others designs include monobloc or deck mixers that are fitted like any other tap but have three handles – the third one diverts cold water via the filter. If you are undecided about how worthwhile it is in your area, buy a cheap filter jug first and try it. If it makes a substantial improvement to the taste it might be worth investing in a filter tap.

**D**

**C**

Kitchen taps are not radically different from bathroom taps, which have already been covered (see page 105). Here we look at the monobloc style of tap, which may also be used in a bathroom.

A monobloc mixer tap must squeeze two supply connections into a single tap body the size of which is limited by the hole in the sink. This rules out 15mm diameter pipes and half-inch pipe threads, such as those on single taps. Monobloc taps have 'tails' instead.

There is no standard for tails, so if you are fitting a new tap consult the fitting instructions. It's a good idea to check before buying your new tap because the tails supplied may not suit your current set up. The simplest type for the manufacturer has a pair of copper tails about 200mm long soldered into the tap body. These are 10mm diameter at the tap end but flare out to 15mm diameter at the other end to suit normal supply side plumbing.

Slightly more upmarket taps have threaded holes in the body that the tails screw into. Some manufacturers include copper tails which, like the soldered versions, flare out to 15mm diameter, and some supply flexible 15mm diameter connectors. If your taps have rigid tails you can use 15mm diameter to 15mm diameter flexible connectors, but that adds to the overall cost and complexity. It's worth checking if flexible tails are available for your preferred tap.

In the sequence a tap with rigid tails is replaced by one with flexible tails so you see both systems in action.

## Replacing a monobloc tap

This is the kitchen that featured on page 102, with the tangled plumbing under the sink.

**1** The first job is to turn off the stopcock, which cuts supply to the cold tap **A**. In this case the house has a combi boiler, so turning off the stopcock cuts supply to the boiler and therefore cuts off the hot water too. In a house with stored water in the loft you need to find an isolation valve, or if there isn't one, drain the tank. Turn on the taps to make sure the water is really off. It will run for a while since the pipes upstairs are all full of water that will drain back down. It's not really necessary to drain the entire system using the drain cock.

**2** The tap tails are just visible behind all the waste pipes but it would be nearly impossible to work on them **B**. Put a bucket or bowl under the waste trap and remove it (see *Unblock a Waste Pipe*, page 118). Then dismantle all the other waste pipes to make room to work. If the system is this complicated make an effort to remember it, or even take a picture, because you are going to have to put it back!

As you will be working with your nose next to the waste pipe it might be a good idea to plug the pipe with tissue first to avoid having to suffer unpleasant smells!

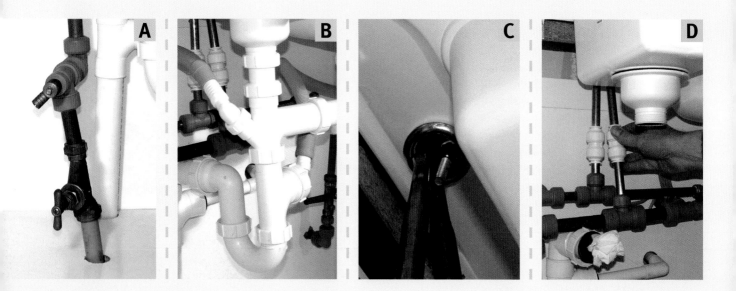

A  B  C  D

**3** Monobloc taps are often held in place by a single stud and nut, and they can be hard to reach. A long box spanner is ideal and plumbers' merchants stock them, but if you are patient you can undo them using an open-end spanner **C**.

**4** The next bit is tricky, as you need to release both joints whilst lifting the tap. This requires at least three hands and considerable dexterity. Compression joints would made things easier **D**. Fortunately there is some flexibility in plumbing, especially when push-fit joints are used, and you may be able to remove one joint at a time. If they are really awkward, lift the tap a little, putting some strain on the joints, then tap each joint downwards alternately by sliding a spanner down the pipes. When they come apart you will get some more water out since the pipes up to the taps are still full.

For comparison both taps have been laid out side by side **E**, though it is actually easier to fit the tap body before attaching the spout, and the compression joints on the flexible tails won't actually pass through the hole in the sink like this. Notice the different fittings too, the newer tap has a much more robust large screw thread and brass nut, and a bronze washer.

**5** To fit the tails, pass them through the sink from below. Slide the rubber washer over them before screwing them into the tap body and tightening them **F**.

**6** Slide the washer over the tails from below, then screw the nut on. Some taps use plastic nuts and washers. If your tap has these, do not over tighten it or you will strip the thread. This one is a sturdy design made in brass **G**.

**7** When the push-fit fittings for the previous tails were removed the pipes were dirty and corroded, so an emery cloth was used to clean them before the compression joints on the new tails were fitted. Make sure the taps are off before you restore the water supply, and keep an eye on the new joints for a while to make sure there are no leaks **H**. Leave the waste pipes off until you are sure everything is OK.

## toptip*

When you first open the taps there will be a lot of air in the pipes, which will cause spluttering and splashing. Open the taps gradually until the flow settles down, then open all the other taps in the house too since air will have got into the feeds to all of them.

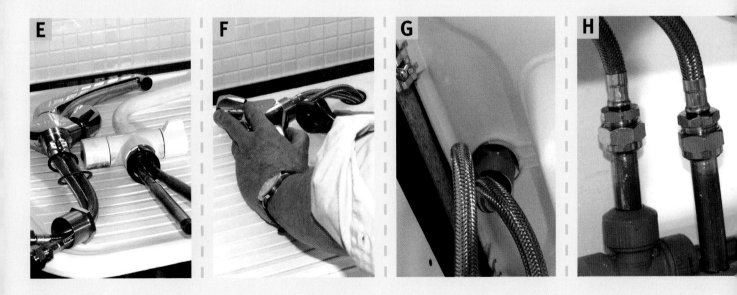

# Heating System Maintenance

There is not space in this book to cover central heating in detail, but the DIY plumber should be able to deal with some of the simpler issues that arise.

## Stuck Radiator Valves

Thermostatic radiator valves allow you to control the temperature of individual rooms, making the system more fuel efficient and your home more comfortable, but they are prone to sticking. If they have been turned down all summer you may find they don't want to open again when you need the heat in the winter. Fortunately they are not hard to fix.

The thermostat is fixed on top of the valve with a nut that should be only hand-tight, though you may need a wrench to shift it if someone has over-tightened it previously. You can safely remove this without any danger of water leaking out.

Underneath the thermostat there is a valve operated by a small pin. When the room temperature matches the thermostat setting the pin is pushed down, cutting off the water flow. When the room cools down, the water pressure should push the pin back up restoring the flow.

Hold the pin with pliers and move it up and down. Don't be rough with it, and take care not to damage it or it will definitely stick in future. A little WD40 sprayed on it might help. After a few movements it should free up so that when you let go it automatically pops up. Press it down and release a few times to make sure it's OK, then screw the thermostat back on.

**REMOVING THE THERMOSTAT**
You can remove the thermostat without disturbing the valve and letting any water out by undoing the thumbscrew (*above*). The pin should pop up automatically if you press it down then release it (*right*).

# do it  BLEED A RADIATOR

Mains water has some air dissolved in it. This is released when the water gets hot, and bubbles to the top of radiators. If a radiator makes a splashing or gurgling noise and the top feels cooler than the middle or bottom, it needs bleeding.

Radiators have a bleed valve fitted in one of the holes near the top (the other is usually fitted with a blanking plug). Most valves have a small square shank that you can turn with a special key. Some valves have a slotted shank, which can be turned with a screwdriver. Have a cloth or handful of paper towel handy and hold it under the valve as you open it **A**. Let the air out and close the valve as soon as water starts to jet out. Beware of hot water if you do this while the heating is on.

New systems will need bleeding several times until they settle down. If you regularly find air in a system there are two possibilities: either there is a leak somewhere and the water dripping out is being replaced by air syphoned in, or more likely it isn't air at all but gas generated inside the system itself.

Radiators are made from steel, which will corrode when wet, and particularly when other metals are attached, such as copper pipes. This electro-corrosion breaks down water into hydrogen and oxygen. The oxygen corrodes the steel, producing a black iron-oxide sludge and eventually causing a leak, and the hydrogen bubbles up to the top of the radiators causing an air lock.

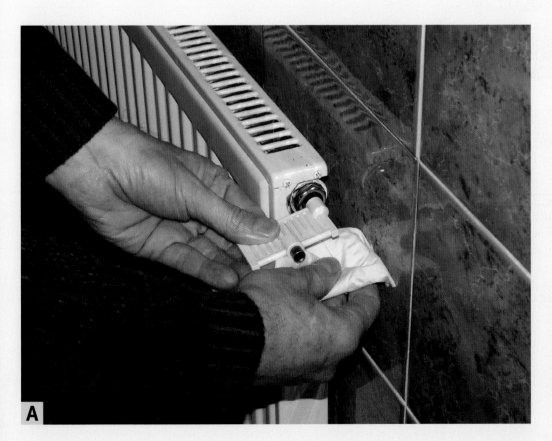

A

**BLEEDING A RADIATOR**
Hold a cloth or paper towel under the valve as you open it.

# doit FLUSH A RADIATOR

If you have a radiator that never seems to get hot, it may be clogged with sludge. This tends to settle at the bottom during the summer when the heating is turned off, impeding the flow of water through the valves. It isn't particularly difficult to flush them out, but it can be messy if you are not careful.

**1** The lockshield valve is usually covered by a cap. Pull this off, and close the valve using a wrench. Close the main valve on the other end, too.

**2** Put old towels or similar around both valves, then slacken them a little . The aim is to tilt the radiator forwards so the joint on the right will tend to undo anyway, whereas the one on the left will tighten, so needs undoing more to start with. Do not undo more than necessary – a few drips of water may escape, but no more.

**3** Lift the radiator about 10mm or so to clear the brackets, then tilt it forwards carefully. Check the joints as you do so, to make sure that they are allowing it to turn without straining the pipework.

**4** Lower the radiator to the floor. At this point the water should be below the height of the valves so you can disconnect it completely. Keep a suitable container handy to empty the water into **B**.

**5** When it is empty, take the radiator outside and flush it out with a hose pipe. You may not think there is a lot in there, but a bit of sludge around the inlet or outlet can make a big difference to the heat output.

**6** To help prevent a repeat, add some central heating protection fluid before you refill the radiator. The normal method is to refit it, remove either the blanking plug or the bleed valve and pour in the fluid using a funnel and hose. If you do not have a funnel, you can pour the fluid into the radiator before refitting it, reversing the procedure used to empty it.

**7** When the radiator is refitted, open both valves, then open the bleed valve to let the air out. If you have a combi boiler you may need to top it up with water (see above). Boilers fed from header tanks top themselves up automatically.

## toptip*

Large radiators are heavy. If you have a steel radiator with a surface area of more than a metre (counting a double panel as double the area) then you will probably need help to support it.

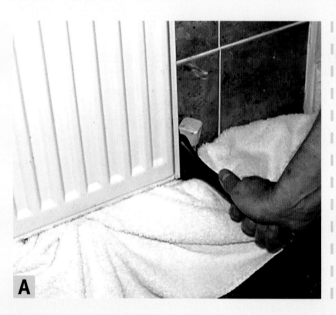

A

B

# REPLACE A RADIATOR

Fitting a new radiator is hardly more complex than removing and replacing an existing one, especially if it is the same size. Whether you are moving the radiator, replacing it with a larger one to increase heat output, or fitting a more stylish one in a different size it involves only basic plumbing and can be done with any of the normal pipe-joining methods. You can even use plastic pipe if you find it easier to handle. If you are going to remove the radiator valves you will need to drain the system, or at least the part of it above the level where you are working. When you refill, don't forget the water treatment.

New radiators normally have threaded holes in each corner. You need to fit 'tails' to two of them, usually the bottom two, a bleed valve to one of the top ones, and a blanking plug to the fourth. It used to be common to have the water entering the radiator at the top and leaving at the bottom on the opposite side. This was to encourage the thermo-syphon effect before circulation pumps were used, but it is rarely seen these days.

There are different types of tails to suit different valves. This radiator has one of each kind **A**. The first has a bulge resembling a compression joint olive formed on the end, and the valve has a matching tapered hole.

Put the nut on the tail before you screw it into the radiator. A few turns of PTFE tape or a little joint sealing compound around the screw will make sure you have a leak-proof joint **B**.

A large Allen key or 'radiator spanner' is needed to tighten the tail. The second type of tail has a plain pipe and is used with valves that have a compression joint fitting. In most cases it makes little difference, but in a tight corner it does allow you to fit an elbow right next to the radiator and place the valve somewhere more convenient. Tails of this type have a hexagonal section so you can tighten them with a spanner **C**.

A

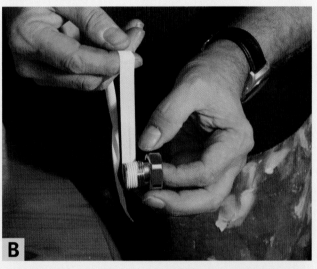

B

C

# 5 Troubleshooting

Plumbing problems tend to arrive out of the blue. One moment everything seems fine; the next you have a damp patch, puddle or worse. It's generally easy to tell where the problem lies, but some problems creep up. It's never a good idea to ignore minor drips and damp, as they won't cure themselves, and may get worse.

# Causes and Cures

## Disturbance

Copper pipework and brass fittings are very robust. Left alone they will survive for 50 years or so, but once you start changing things by fitting new taps or replacing the sink unit you run the risk of disturbing joints that were maybe not well made in the first place. If you do any work on an old system, keep an eye on the surrounding joints and fittings for few days until you are confident you haven't disturbed anything.

## Corrosion

Copper and brass corrode, but so slowly that they don't normally cause leaks. Corrosion around rarely-used valves and stop cocks can make them hard to turn. It is a good idea to try turning them off and on occasionally to ensure they will work when needed.

Steel radiators are much worse (see page 109). If the seams look rusty you have a problem looming, and if one starts to leak there is a very good chance the rest will follow shortly, so you need to start planning to replace them all. When you do, make sure you add radiator protection to slow down future corrosion.

Plastic doesn't corrode like metal, but it does degrade over time. Plastic plumbing for water supply is relatively new but has been formulated to last 25 years. There was a problem in the US during the 1990s with poor quality plastic fittings, but EU and UK standards have been defined to avoid a repeat here.

Plastic waste fittings have been around much longer and early PVC was not UV stable. Where waste pipes went out through the wall into daylight, they should have been painted to protect them, but many were not. Such pipes become brittle and crack easily if something bumps into them. Internal waste pipes are usually OK because they don't see much daylight. Modern waste pipes are usually UPVC (or PVC-U), which is resistant to the effects of UV light.

**RUSTY RAD**
When a steel radiator starts to rust it will only be a matter of time before it leaks.

## Pressure

A joint that was only just good enough at a low pressure will start to leak if there is an increase in water pressure. This can happen if you change from an indirect to a direct system but retain some of the original pipework, so do check the whole house when you first switch on the new system.

Central heating systems that are served by comb boilers are usually pressurised to around 1.5 Bar. These types of boilers have a pressure gauge that shows the current situation and a pressure release valve that will open automatically if the boiler is over-pressured. In normal operation the pressure will rise a little when the boiler is on and fall when it cools, but problems can arise.

In order to fill the system when it is installed there is a temporary connection, usually made with a flexible connector, via an isolation valve. Once the system is filled and tested the engineer will remove the connection to prevent any chance of contaminated water from the central heating system entering the supply. At times, you will need to bleed the radiators and this can lower the pressure. If the pressure drops too low you cannot bleed all the gas out and the boiler may close down to prevent damage. This means that you will need to reconnect the supply and top it up.

If this happens, there are a couple of things that can go wrong. Firstly, it does not take much water to top up a system, and if you overdo it you may find that some joints will start to leak before the pressure release valve lifts. Secondly, it's tempting to leave the flexible coupling attached and rely on the isolation valve. If, however, this is even slightly faulty the pressure will rise very gradually, perhaps over months, until eventually it either opens the safety valve or causes a leak somewhere. Don't leave it connected!

## doit MEASURING PRESSURE

A bar is a unit of pressure roughly equal to normal atmospheric pressure. When used on pressure gauges such as boilers, the gauge measures the pressure difference that is between the atmosphere and the inside of the boiler – so 1 bar on the gauge, which is sometimes given as 1 bar(g), is actually approximately 2 bar.

In the photograph the pressure gauge is on the far right and the time clock for the system is on the right.

## Blocked Waste Pipes

We put some unpleasant things down our waste pipes and generally they just take it all away. Occasional blockages caused by misuse are just a fact of life, but if your waste system blocks regularly it was either badly designed and installed in the first place, or something has gone wrong with it.

**PIPE PROBLEMS**
Freezing water pipes can burst pipes or force joints apart like this (*above*). Blockages to waste pipes are fairly common, and sharp bends (*above right*) are more likely to clog. Two 135° bends like this are better than 90°, especially for macerator waste.

## Frost Damage

Water has a unique property: it expands below 4°C, and expands even more when it freezes. If water freezes in a pipe something has to give. If the pipe is smooth and leads to a system with some expansion room it may just slide along, more commonly it gets stuck in elbows, valves and taps then bursts a pipe or forces a joint apart.

The photograph above, left shows what happened when the householder (the author) remembered to turn off the isolation valve leading to the outside tap before a particularly severe frost, but forgot to open the tap to let the water out. Hydraulic pressure has forced the olive right off the pipe. Fortunately, a smear of jointing compound and a new olive will fix it.

## Poor Design

Waste pipes should always 'fall' – that is slope downwards – in the direction you want the waste to go. About 20–25mm in every metre is enough. If you have sections that are level or slope the wrong way then solids in the waste get the chance to settle and stick to the pipe, eventually clogging it up.

Bends are an issue too. Waste water only falls a small distance from the fitting to the stack or drain, so doesn't have a great deal of pressure to push it along. Sudden changes in direction slow down the flow, which may cause solids to be deposited in the pipe. If a particular elbow or tee is prone to clogging, try replacing it with a more gentle curve. Two 135° elbows rather than one 90° or a Y joint and a single 135° elbow rather than a conventional tee will speed up the flow. This is particularly useful for waste plumbing that has to carry more than the average amount of solid matter.

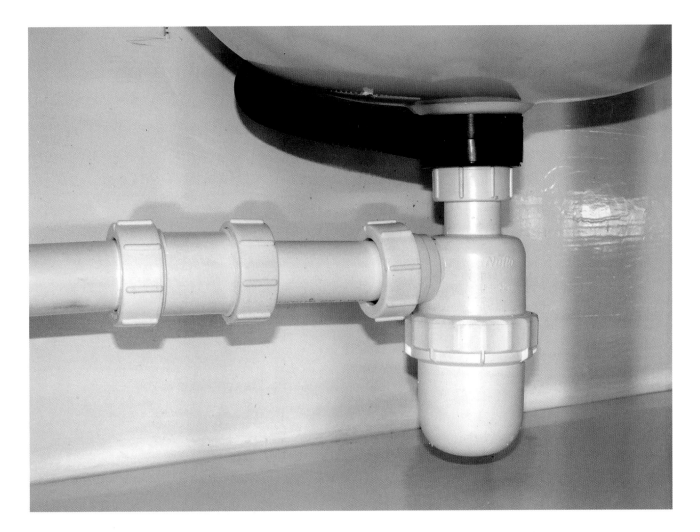

**FALLING WASTE**
Waste pipes should slope down (fall) at least 20mm per metre. This level section will be prone to blocking, but at least the joints make it easy to clear.

Changes in diameter are also bad news. It is OK to connect a smaller waste pipe into a larger one, but connecting, say, a Ø40 pipe from a bath to a Ø30 pipe previously used by a wash basin is likely to lead to a blockage.

Combining waste pipes from a basin and bath or bath and shower is common practice. Take care: if the run from the junction to the stack (or gulley) is more than a metre or so, it may create a syphon effect, and letting the water out of the bath will empty the other traps and cause unpleasant smells. If it is impractical to re-fit the pipework, fit anti-syphon traps (see page 65). Do not combine pipes from different floors, or you bathwater may re-appear in the kitchen sink!

## Faults

Correctly designed and fitted waste systems do not go wrong, but they can be maltreated. Refurbishing buildings, especially if it involves changing the use of rooms, can lead to all manner of abuse. If you are fitting a new kitchen or bathroom, adding a downstairs WC or an en-suite shower, take care not to damage the existing waste system, or to overload it with new fittings.

Even disconnecting something can cause problems if not done properly. In one example, a kitchen was being converted and no longer needed a waste pipe. The builder sawed the pipe off level with the floor, hammered it down a bit, then concreted over it. A few weeks later all the WCs overflowed. The waste pipe from the kitchen had connected to a drain under the floor. When the builder hammered it down it went through the drain, partially blocking it. Two weeks' worth of toilet paper then clogged up the narrowed space. The builder had to dig up the concrete floor, replace the drain then re-instate the floor – not something a DIY plumber would really relish.

# do it UNBLOCK A WASTE PIPE

Pipes tend to block at elbows and junctions, and the most common blockage of all is in the trap. All modern traps are designed to be easy to open and clear.

**1** The bottom section of this P trap under a kitchen sink can be removed by hand **A**. Place a bucket or bowl underneath it to catch any fall out. A waste system like this is made of many individual parts, each of which can be removed for cleaning if necessary.

**2** The bottom section of a bottle trap beneath a basin can be unscrewed for clearing **B**.

**3** A bath trap usually has an access hole for clearing **C**. Remove or unscrew the cover, then poke the trap clear with a piece of wire.

**4** A shower trap is designed to be opened from above **D**.

**5** If the pipe beyond the trap is blocked you might try drain-clearing chemicals. The professional versions are much stronger than those found in supermarkets and need to be handled with care; gloves and goggles should be worn. Pouring these chemicals into a trap full of dirty water will dilute them before they even reach the blockage. They will work far better if you empty the trap, replace it and then pour in enough fluid to fill the trap and run down the pipe. These chemicals can damage the plating on the waste outlet, so flush round it with a small amount of water immediately after pouring the chemical down.

**6** Leave the chemical for 15 minutes or so (or as specified in the instructions), then flush with water.

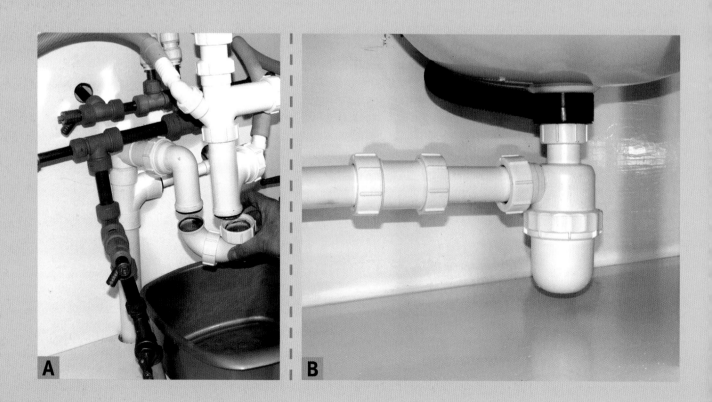

A

B

## UNBLOCKING DRAINS

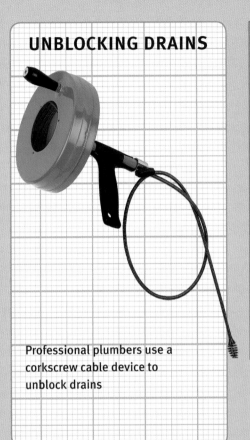

Professional plumbers use a corkscrew cable device to unblock drains

**CAUTION**
## CHEMICALS

If pouring drain-clearing chemicals into a blocked pipe does not work, you may have to remove a trap or disconnect some joints to clear the blockage. If you do this, remember that the pipe will be full of dangerous chemicals. Wear gloves and goggles and take great care.

**7** If chemicals don't work, or you don't fancy using them, dismantle the trap again and try poking down the waste pipe with wire. Plumbers use a corkscrew cable device like the one shown left. It has a small spiral tool on the end to penetrate the blockage connected to a flexible drive cable. Turning the handle spins the cable. Not all DIY plumbers will have this, but a flexible curtain wire of the type used for net curtains, or even a piece of lighting cable (not flex) will often work.

**8** If you cannot clear the blockage from there, try the other end. If the waste pipe runs into a stack you might find an access port in the soil stack opposite the junction, so poke the wire up from there. Beware of the splash back if you succeed, and put the WCs out of bounds to the rest of the household while you work on the stack!

C

D

# Glossary

**Air lock**
Air that is trapped in a pipe, and is reducing or even preventing the flow of water.

**Atmospheric pressure**
The pressure in the atmosphere caused by the force of gravity.

**Allen key**
A hexagonal tool that is designed to fit into and turn socketed screws.

**Back flow**
Flow in a direction contrary to the natural or intended direction.

**Bar**
Unit used to measure pressure. 1 Bar is roughly normal atmospheric pressure – what meteorologists call 1000 millibars.

**Bleed valve**
A valve designed to release trapped air in a central heating system.

**Butane**
A gas that is present in natural gas, and is sold compressed in cylinders for domestic or industrial use

**Ceramic**
Fired pottery.

**Check valve**
A device used to protect the water supply from contaminants.

**Cistern**
Literally a tank. The term is usually applied to a tank for flushing a WC but is sometimes used for a tank for storing water in indirect systems.

**Close-coupled**
A term used to describe a WC suite with a cistern that is bolted or clamped to a projection on the upper rear edge of the pan.

**Combi (combination) boiler**
A combined DHW and central heating boiler that heats water on demand.

**Compression rings**
Fittings that seal by squeezing a ring (olive) onto the pipe.

**Conex**
Brand name for widely-used compression fittings.

**Corrosion**
Degradation of a material caused by air, water, acids, alkalis, chemicals or electrolytic action.

**Coupling**
A straight fitting for connecting plastic or copper pipes in line.

**Cuprofit**
Brand name for commonly-used copper pushfit fittings.

**Deck (taps)**
Wide bodied mixer taps to fit two hole baths, basins or sinks.

**Direct system**
A heating and DHW system that runs on mains water pressure.

**DHW**
Domestic hot water – plumber's shorthand for anything to do with the hot water system in a house.

**Drain valve**
A valve fitted at the lowest point of a water system or section of pipework that is not self-draining, which allows the sytem to be emptied.

**Elbow**
A joint with a sharp corenr or change of direction, usually 90° or 135°.

**Expansion joint**
A joint that is fitted in pipework to allow for the linear expansion of the pipe material when the water temperature is raised.

**Flux**
Chemical paste that is applied to copper pipe when making a soldered joint. It prevents copper oxidising when it is heated for soldering.

**Furring**
Encrustations of hard water lime or other salts deposited by the water that is heated in a pipe or appliance

**Gate valve**
A shut-off valve used on low-pressure water pipes, such as the outlet pipes from a water tank.

### Hot water cylinder
An insulated tank to store domestic hot water in an indirect system.

### Immersion heater
An electrical heater that fits inside a hot water cylinder.

### Indirect system
A heating and DHW system that is gravity fed from a tank, usually situated in the loft.

### Immersion heater
An electrical element fitted in a hot water storage vessel and controlled by a thermostat and valve.

### Isolating valve
Fitting used to stop the flow of water to a section of pipe or an appliance.

### Kite mark
The British Standard mark placed on items manufactured to a approved standard; a guarantee of quality.

### Localised hot-water system
A system that heats water at the point at which it is required.

### Mait
A type of jointing compound. Not to be confused with a plumber's mate.

### Multi-point water heater
A water heater, usually gas, that supplies hot water to several taps.

### Nominal size
Convenient designation for the size of a pipe, fitting or other component, generally a rounded-up number.

### Olive
A copper or brass ring that is used in a compression fitting.

### Open-vented
A system incorporating a vent pipe that is open to the atmosphere.

### Overflow pipe
A pipe designed to discharge excess water safely.

### Potable water
Wholesome water that is fit for human consumption.

### Power shower
A shower system incorporating a pump to boost the water pressure.

### PTFE
Poly(tetrafluoroethane). Used in tape form to seal joints.

### Push-fit
A plumbing fitting designed to create a watertight joint simply by inserting a pipe.

### Relief valve
A valve used to limit the maximum pressure in a system. It will open automatically if the desired pressure is exceeded, allowing fluid to escape.

### Rodding
Method of clearing obstructions or blockages from drains or sewers by using connecting, flexible rods that are pushed into the pipeline.

### Rising main
The pipe that conveys fresh water from the incoming supply to the storage tank in the loft .

### Silicone sealant
A sealing compound often applied round sinks or between a worktop and wall. It cures without hardening to form a water-repellent joint.

### Single-stack system
A form of one-pipe system that may have waste water and soil discharging into it. All or most of the trap ventilating pipes are omitted.

### Soil pipe
A pipe that conveys the discharge from a WC or urinal.

### Soil stack
Large vertical pipe, usually 110 mm diameter taking soil and waste water to a drain.

### Solder
An alloy of tin and copper (or antimony) used to join pipes. Before 2006 it contained lead which is now banned in the EU.

### Solder-ring

Copper fittings with a ring of solder embedded inside.

### Stainless steel

A steel containing chromium and nickel that is highly resistant to corrosion.

### Standpipe

A vertical waste pipe fitted with a U-trap that is used to insert the flexible waste hose from a washing machine or a dishwasher.

### Stopcock

A shut-off valve used on high-pressure water pipes such as an incoming water supply.

### Storage hot-water system

Stored heated water ready to be supplied to one or more outlets.

### Tee

Pipe fitting in the shape of a T used to joint three pipes together. Pipes can be the same size or different sizes (equal / unequal tee).

### Trap

A fitting under a waste outlet that is designed to hold water and prevent drain odours entering the house.

### Uni-compression

Versatile compression joints used for waste systems. Not quite 'universal' but will tolerate some variation in pipe size.

### Valve

Device to open or close a flow of liquid, gas or air, or to regulate a flow

### Waste water

Discharges from appliances that do not contain human or animal excrement, such as basins, sinks, baths and shower trays.

### Water line

A line marked inside a cistern that indicates the highest level of water at which the supply valve should be adjusted to shut off the supply.

### WC

Water Closet, toilet, lavatory, loo, bog, john, throne – or any number of other names. We used WC in the book because it's shorter!

### Yorkshire

Brand name of common solder-ring joints.

# Suppliers

**B&Q**
www.diy.com

**Cifial**
www.cifial.co.uk

**Conex Cuprofit**
www.ibpconex.co.uk

**Hepworth**
www.hepworthplumbing.co.uk

**Homebase**
www.homebase.co.uk

**John Guest**
www.johnguest.com

**Marley**
www.marleyplumbinganddrainage.com

**Mira**
www.mirashowersales.co.uk

**Plumb Centre**
www.wolseley.co.uk

**Screwfix**
www.screwfix.com

**Watkin and Williams**
www.watkinwilliams.co.uk

**Wickes**
www.wickes.co.uk

The publisher is not responsible for the content of external internet sites.

# About the Author

Phil Thane has had many jobs, including design technology teacher, software support manager and writer. All have involved explaining technical matters in as clear and concise a manner as possible.

A keen DIY-er, he re-plumbed his first house and installed central heating. Now living in North Wales, he is gradually creating a terraced garden on a neglected hillside.

## Acknowledgements

All photography by Phil Thane with the following exceptions:

Cifial UK Ltd Cifial (www.cifial.co.uk):8; 14T; 105 (T&M)

Conex (www.ibpconex.co.uk): 40TR; 41B; 42; 44BR

Flickr: axeldeviaje: 87T; exfordy: 4;  GeS: 1BL, 52TR; 52B;
spierzchalala: 11M; scalespeeder: 69M; tanais: 76; krikit: 78TR

Hepworth (www.hepworthplumbing.co.uk): 33, 34T; 35; 37B; 43T; 45B

Alison Howard: 52TL

Mira Ltd Mira (www.mirashowersales.co.uk): 1 (TR, TL&BR); 2, 7B; 9; 98; 100; 112

Gilda Pacitti: 29 TR

PWS Distributors Ltd: 5; 6T; 10; 11L; 16;

Screwfix (www.screwfix.com): 13T; 14B; 23T; 23 (L&R); 24 (S&R); 26 (T&B); 27 (all images); 28T; 29 (all spanners); 32 (TL&TR; 38BR; 39BR; 50M; 56 (all images); 57 (M, BL&BR); 61TR; 62; 63 (TL&TR); 64 (BL&BR); 65T; 73M; 80T; 80B; 84; 89 (L&R); 91B; 101; 104; 105B; 119TL

Stock xchng: (remind): 6M, 6B, 7T; 68; 78;

T=top, M=middle, B=bottom, L=left, R=right, S=second, TH=third
Note: every effort has been made to ensure the accuracy of this list. If you have any queries, please contact GMC at the address on page 128

# Useful Websites

The Chartered Institute of Plumbing and Heating Engineering (CIPHE) is the UK's professional and technical body for plumbing and heating professionals. www.ciphe.org.uk.

Plumbing Pages has information on courses and routes to professional plumbing qualifications. See also British Plumbers Employers Council: www.bpec.org.uk

## Gas

For information relating to gas safety: www.gassaferegister.co.uk

## Electrical work

The National Inspection Council for Electrical Installation Contracting (NICEIC) is the body that regulates the professionals. Its website has useful information for amateurs too, and publishes detailed guides: www.niceic.org.uk

The Electrical Safety Council offers a wealth of useful information in its guidance notes: www.electricalsafetycouncil.org.uk

Information about electrical testing can be found at: www.electrical-testing-safety.co.uk

For information on electrical safety regulations log on to: www/planningportal.gov.uk

For information on the Electrical Contractors Association log on to: www.eca.co.uk

## General DIY advice

Advice including details of current UK legislation relating to plumbing work: www.diyfixit.co.uk www.completediyguide.co.uk www.diynot.com.

## Energy saving

For information on recommended energy saving products, see: www.energysavingtrust.org.uk

The publisher is not responsible for the content of external internet sites.

# Index

To request a full catalogue of GMC titles, please contact:
GMC Publications
Castle Place
166 High Street
Lewes
East Sussex
BN7 1XU
United Kingdom

Tel: 01273 488005
Fax: 01273 402866

Website: www.thegmcgroup.com